ABOUT ROTHKO

Frontispiece   Mark Rothko in his studio, 1952

ABOUT ROTHKO

DORE ASHTON

New York
OXFORD UNIVERSITY PRESS
1983

Copyright © 1983 by Oxford University Press, Inc.
First published by Oxford University Press, New York, 1983
First issued as an Oxford University Press paperback, 1983

Library of Congress Cataloging in Publication Data
Ashton, Dore.
About Rothko.
Bibliography: p   Includes index.
1. Rothko, Mark, 1903-1970.
2. Painters—United States—Biography.   I. Title.
ND237.R725A93   1983   759.13 [B]   83-2268
ISBN 0-19-503348-5   ISBN 0-19-503349-3 (pbk.)

Printing (last digit): 9 8 7 6 5 4 3 2 1

Printed in the United States of America

PREFACE

I call this book *About Rothko* because in both senses—all around
and approximately—writing of painting is always "about." Even
more, writing of an artist is always approximate. No amount of fac-
tual or biographical detail can ever comprise an adequate portrait.
All biographers know how maddeningly diverse responses to an
artist's personality can be. I have had countless experiences with
Rothko's friends and acquaintances that bear out my own perception that Rothko's complex personality yields only rarely to gen-
eralizations. His nature revealed itself reluctantly, even to friends,
and they often had totally opposite reactions. To a large degree I
have finally relied on my own friendship with Rothko that lasted
from my youth to his death in 1970—some eighteen years.

Most of my conversations with Rothko occurred during studio
visits or over lunch. In the early years it was mostly pastrami sandwiches in 6th Avenue delicatessens. Toward the end, we lunched in
elegant uptown Chinese restaurants. But the conversations were
much the same. We fell into a kind of philosophical badinage. We
talked about "life" the way we imagined the old Russian literati
used to talk—long, meandering conversations that often concluded
with a sigh, and an agreement that it is hard to find a way to live
an ethical life. Most of our allusions to art were within that larger
context.

When I first knew him, I caught glimpses of the wit and gusto for which Rothko was known in his earlier years. There was often tenderness in his greeting, sometimes even exuberance. Many friends recall the great warmth of his salutational smile, his eyes soft behind his thick-lensed glasses. Sometimes after the first moments a certain distractedness descended. Rothko was a nervous and often impatient man. Sometimes at dinner in my home, he would get up and wander between courses, cigarette in hand. But there were also times when he was intensely interested in a certain subject and would speak with great animation and listen intently. When his mind was distracted, his shambling gait and a certain vagueness in his glance would indicate his unwillingness to linger in conversation.

There was certainly anger in Rothko. Most of his close friends accepted it and understood that some of it was bred in the loneliness of the studio routine, and some evoked by his general condition of doubt. Like K. in *The Castle*, Rothko spoke the same language as the others, but for some permanently obscured reasons, the others didn't often understand. Perhaps his *pudeur*, which prevented him from exhibiting what was certainly a romantic and sometimes sentimental soul, contributed to his isolation. He spent a lot of emotional energy preserving himself from intrusion and fending off overheated responses to his life's work. Although his paintings progressed from metamorphosis to transfiguration, he adamantly denied in public that they represented for him (and now for many others) a kind of *ekstasis* (*ex:* out of; *histanai:* to set, to stand). Yet, he was most affectionate with those who sensed the depth of his passion, and who knew that his protestations were somehow ritualistic.

Rothko committed suicide February 25, 1970. He left an estate in such disarray that one of the most complicated lawsuits ever to take place after the death of an artist sensationalized his name. Since I am writing about what Rothko called "the meaning of a man's life's work," there is no place in my book for the notorious trial.

New York   D. A.
December 1982

ACKNOWLEDGMENTS

I thank Bonnie Clearwater of The Rothko Foundation, and the Archives of American Art for their generous and patient assistance. Many friends and acquaintances have shared their experiences with me, and they all have my warmest gratitude: Andrew Forge, George Dennison, Ronald Christ, Matti Megged, Ned O'Gorman, Robert Motherwell, Richard Roud, Carla Panicali, Carlo Battaglia, Toti Scialoja, Gabriella Drudi, Elaine de Kooning, Sir Norman Reid, Peter Selz, Herbert Ferber, Donald Blinken, Joseph Liss and Mrs. H. R. Hays. A special thanks to Tony Outhwaite, whose enthusiasm spurred my work; to Geoffrey Hoffeld of the Pace Gallery, and above all to Kate Rothko Prizel for her graciousness.

D. A.

ABOUT ROTHKO

Ah, whom then can we prevail upon in our need? Not angels, not men,
and the clever animals see at once
that we are not very securely at home
in the interpreted world.

RILKE, "First Duino Elegy,"
Translated by Edward Snow

1

Rothko did not feel "very securely at home in the interpreted world." He looked about him. He searched faces. He traveled. He married and had children. But he was never at ease. He was indifferent to objects, and took little pleasure in the ordinary embellishments of daily life. In his various studios austerity reigned. There were no distractions—no bibelots, reproductions, Oriental rugs or even comfortable chairs. His sensuous and emotional life were not dependent on paraphernalia or possessions, even small ones. He craved transport, and found it mostly in music. His greatest fund of emotion was lavished primarily on what he made—paintings. Those paintings were to be his passport to a more luminous world, not encumbered by our nouns and adjectives, our interpretations that always fall short. They were prepared by careful thought, nurtured by well-fondled ideas, but, as he said, "Ideas and plans that existed in the mind at the start were simply the doorway through which one left the world in which they occur."[1] To leave the world in which ideas and plans—so quickly superseded by emotions—occur was essential to Rothko. The doorway was as eagerly sought as Kafka's portal. His was a temperament that had unquenchable thirst and hunger. He had deep needs to fulfill, many of them incapable of being brought to the threshold of language.

Like many artists of the modern epoch, Rothko could not make use of either angels or men. The angelic order, he sensed, was no longer available. He was a Nietzsche admirer. But the world of men, interpreted, as it was, always by its surfaces, could not be enough. Although he made a *salto mortale* in his paintings, and was immensely admired as well as reviled for his departure to the pure realm, he resolutely resisted attempts to thrust him into categories that bordered on mysticism. He hovered, in his mind, on the threshold of the doorway, but was wary, like most moderns, of an irreversible journey. He insisted he was a materialist: "When I wrote the introduction to Clyfford Still's catalogue for Peggy Guggenheim, I spoke of earth worms (primordial matter)—that's far from *otherworldliness!*"[2]

And yet, there were the works—his own environment wrought over a long period that, to so many, spoke of another place. There were countless responses to the gesture of renunciation that brought Rothko into a realm in which his needs could be fulfilled, at least partially. The large abstract paintings that moved so many, and continue to move those attuned to such experiences, could not be perceived as the work of a "materialist," and there were times when Rothko himself alluded to their "transcendence." Somewhere in him was the lover of the absolute. Some viewers responded immediately to the experience; others were skeptical. Those who were not emotionally available, or who saw sameness in Rothko's later work, with its simple and repetitive compositional schemes, were irritated by and, at times, contemptuous of the extravagant claims made by those who were, often unaccountably, moved. The derogators harped upon the sameness. Brian O'Doherty insisted that "Rothko's art utters a single word, insistently." Rothko's defenders found little to say that could convert the unbelievers. It was quite true, as O'Doherty and others complained, that the literature on Rothko's work tended toward the exalted, and was always "subjective." The analysis of individual paintings, on which detractors could build their case, seemed a secondary concern to those who were moved and wished to bear witness to the power of Rothko's oeuvre on an unnameable spiritual or emotional level. They—the ones who spoke of sinking into his spaces; of feeling lighter, or of feeling awe; of having a spiritual experience—had understood that Rothko's "enterprise," as he referred to his life's work, was to find some philosophical resting place. Like many poets and philosophers, he needed to feel (not only think) some kind of underlying unity in the havoc of existence. Willem de Kooning understood very well. He congratulated Rothko on his exhibition at the Museum of Modern Art in 1961, saying "Your house has many mansions." He explained: "You see, because knowing the way he made them, all those paintings became like one with many mansions."

Certainly it was often a house divided. Rothko was an intellectually restless man. He was by nature a debater, first of all with himself. Those who noticed his contradictions, both in his few public statements, and in his concourse with others, sometimes remarked that Rothko could be like a lawyer, or like a devil's advocate. Yet he seemed to long for the release that would reveal a wholeness beneath what seemed to him, and so many other modern creators, a devastatingly chaotic existence.

If he talked again and again of his "enterprise" it was to emphasize to himself the potential meaning of its wholeness. On occasion he would call it an "ellipse." Enterprise: "An undertaking, esp. one which involves courage, energy or the like; an important or daring project; a venture" (Webster). Rothko's "project" went beyond painting, although painting, finally, was his only means.

A friend of Rothko's youth, the painter Joseph Solman, marvelled at what he called Rothko's "growth." Rothko was after all forty-six years old in 1949, the year that he made his most daring painting gesture. All the years before he had struggled to make painting the instrument of his inner life, often with discouraging results. Yet something had prepared him for his mature decisions. His insecurity in the interpreted world was probably the deepest force, driving him past its barriers throughout his life. Such uneasiness is a familiar condition of certain artists. Rothko's older brother stressed that he had been a highstrung, noticeably sensitive child. In addition to an inborn uneasiness, however, there were circumstances in his early childhood that worked against an easy adaptation to vicissitudes.

In a life with many turnings, there are fragments of memory that return and retreat. Certain fixed images that have somehow withstood the avalanche of events that overtake a man, seem to have a bearing on his evolution. Rothko carried in him childhood impressions that were ineradicable. He remembered his early family life without enthusiasm, and it would be hard to tell how it affected his life. But the external events that affected his family were deeply graven. When Rothko reminisced about his early childhood he most often recalled the situation, *his* situation, as a Jew among hostile Russians. He did it obliquely: he told of being a child of five and watching, in terror, as Cossacks brandishing their *nagaikas* bore down on him. Those eager to disparage Rothko's self-dramatization are quick to point out that he was born in Dvinsk, one of the most comfortable cities for Jews at the turn of the century. On the whole, the large Jewish population of Dvinsk was spared the atrocious pogroms that took place in other towns in the Jewish Pale, although they shared in the frequent indignities all Jews in Czarist Russia endured. Dvinsk before the First World War had a population of roughly 90,000, nearly half of whom were Jews. It was a busy railroad junction; an important military fortress, and an unusually developed industrial town. Labor unrest was constant after the turn of the century. Many of the burgeoning radical political movements were well represented in Dvinsk, including the newly formed Zionist socialist groups. After the failure of the 1905 revolution, Dvinsk was carefully watched by the Czar's secret police. When the Cossacks came, often to break strikes mercilessly, the Jews were their first target. During the year of Rothko's birth, 1903, there had been the terrible three-day pogrom in Kishinev, followed by others in Russian towns with increasing frequency. Jewish communities after 1903 lived in constant fear, as slogans such as "Destroy the Jews and Save Russia!" were increasingly sanctioned by the Czar. Old friends have remained skeptical that Rothko ever encountered the rampaging Cossacks. Yet, of all the possible memories he could preserve of his first ten years, that is the memory that he most often cited. It could stand for everything else. Certainly the position of the Rothkowitz family in 1903 was precarious. Every Jew knew the dangers. Although Rothko's father, Jacob, was a professional and made a comfortable living as a pharmacist, life for the Jews was never without fear. When Rothko was two years old the first revolution broke upon Russia, driving the Czar to further extremes. In

6

Dvinsk, where there was an educated Jewish populace and where progressive political views were common, the anxiety mounted and surely conveyed itself to the youngest child.

Although Rothko came from a family of emancipated Jews (his parents were ardent Zionists and spoke Hebrew as well as Russian),[3] his father decided that his youngest son would have a formal religious education. The child was sent, probably at the age of five, to a Cheder, a religious school, where in all probability he began the process of memorizing the Talmud and learning the Hebrew scriptures under the strict eyes of teachers who had little thought for the individual differences among their students. Even if Rothko attended one of the more enlightened religious schools, where instruction was sometimes conducted in Yiddish, and where there were the beginnings of a liberal education, the strictness of the regime, the demands on young scholars, and the emphasis on the authority of the teacher could have certainly contributed to Rothko's later hatred of invested authority. At home there were three siblings, all much older: Sonya, fourteen, Moise, eleven, and Albert eight years, when Rothko was born. There was not much latitude for play. Rothko often remarked that he had never had a chance to be a child.

When Rothko was seven years old, his father decided to make the break that so many Russian Jews had made during the twenty years of turmoil. He left for America where a brother had already established himself, planning to bring his family as soon as he could. Within a year he was able to arrange for the escape of Albert and Moise, who were nearing the age of conscription—a disaster that all Jews dreaded. Young Marcus was left with his mother and older sister in a community in general upheaval. Among their middle-class acquaintances there were undoubtedly young radicals who were quick to predict the coming social revolution and whose efforts were frequently terminated by the increasingly violent methods of the Czar's secret police. An intelligent boy nearing the age of ten would not have failed to respond to the extremely volatile situation. The knout of the Cossack was the symbol that summed up Rothko's memories of his Russian childhood.

When, in 1913, Jacob finally sent for the rest of his family, they were able to travel second class, escaping the agony of so many Russian Jews whose experiences in steerage are legendary. They arrived knowing no English and undertook their long train journey to Portland, Oregon, where the father had found employment in his brother's clothing business. Rothko's memories of his life in Portland were tinged with the bitterness of the poor relative. His father had died within months of their arrival, and Rothko's mother, sister, and older brothers were faced with the immediate problems of survival. They were helped, but apparently grudgingly, by the already established family who employed the older boys in menial positions, and put young Marcus to work in the stockroom at a very early age. A life of selling newspapers and working regularly after school did not undermine Rothko. It seemed to enhance his underlying seriousness. He quickly mastered English and while still in grade school became deeply interested in the social problems discussed just as avidly as they had been in Russia. Portland's Jews lived, as they had in their mother country, largely in one neighborhood, and maintained their ties through sharing a common language, Yiddish or Russian, and congregating in community centers. Families such as the Rothkowitzes had already been broadly exposed to theories of social liberation in Russia. In America, they experienced exhilarating relief from the perils of an authoritarian regime, and they were quick to respond to native radical movements. At an early age Rothko found an outlet for his passionate temperament in the excited discussions of social radicalism that took place in the Jewish community center. He even developed a reputation as a skilled debater, according to his brother. He always remembered his attraction to radicalism with an affection for the child he had been—eagerly attending the mass meetings at which such colorful anarchists as Emma Goldman, William Haywood, and assorted figures in the Industrial Workers of the World harangued on various issues, from the right to strike to birth control. Portland was an important stop on the West Coast circuit. Long before the Rothkowitzes arrived it had harbored a small community of intellectuals eager to preserve constitutional rights. The extravagant child of the anarchist movement, Emma Goldman, had early established her name there as one whose rights to free speech must be protected. In her memoirs she warmly recalls certain distinguished citizens of Portland, among them an ex-senator, a Unitarian minister, and an editor of the *Oregonian,* all of whom insisted on her right of free speech. The I.W.W. held frequent meetings in Portland, bringing the workers messages from the East and Midwest where a number of strikes had resulted in almost military skirmishes. The workers were most often the casualties. Pinkerton men, hired by the big capitalists, were never loath to finish off the militants, and where they failed, the U.S. courts often succeeded. The undertone of violence in American life and the outrages obviously bestirred the young immigrants who were all too familiar with social injustice. While no child could avoid wishing to leave his foreignness behind, many of the Russian Jewish youths clung to their social revolutionary theories, carried with them from childhood. Rothko was no exception. Often in later years, he would stress his Russianness. He wished to be perceived as Russian.

Rothko was fourteen when the Russian Revolution occurred, bringing great excitement to the young American radicals and deeply marking American life. Not only did the revolution fire American social idealists with new zeal, but it frightened the government, as it did other governments. After any large-scale social upheaval, most governments, as Emma Goldman was quick to point out, resort to repression. The Americans were no exception. For the youths of Rothko's generation, it seemed essential to fight reaction and to stand for the working man's rights. Rothko threw himself into the spirit and dreamed of being a labor leader. Certainly when he graduated from Lincoln High School in only three years with what the *Oregonian* described as a brilliant record, his artistic yearnings had not yet focused. He had already taken some art classes, and shown a serious interest in music (he taught himself to play the mandolin), but the expectations of the immigrant Jewish community did not lie in the arts. When he and two other Portland Russian-Jewish immigrants were granted scholarships by Yale University, Rothko went with the intention of finding a respectable profession—the kind that his background and formation would honor.

If Princeton in the 1920s was as snobbish as F. Scott Fitzgerald reported, Yale was even more so. It was an unlikely place for three young radicals. Rothko could not have been comfortable within the ivy-clad halls where it was well known that there was a Jewish quota, and where no Jew had ever been appointed to the permanent faculty. He apparently spent a lot of time off campus with his New Haven relatives, some of whom remember intense discussions with him about the arts, particularly music. He followed the usual courses in liberal arts and sciences; was good at mathematics, and in his second year when his scholarship was terminated, had to work as a waiter and delivery boy to make ends meet. His interest in social issues did not abate, and was reflected in the fact that he and two others published, briefly, a critical newsletter the response to which may have contributed to his total disaffection with Yale and his decision to quit after his sophomore year. One of his fellow students remembers that he hardly seemed to study, but that he was a voracious reader. If he did, indeed, see himself in the Russian tradition of youthful radicalism he would have found the classroom tedious, and prided himself on his education through his own reading. He clearly admired the anarchists who, at their best, were readers in several languages. Emma Goldman fairly often gave lectures on the Russian writers and also on the German philosopher Nietzsche. With his thoughts on "one big union" as Haywood used to call the I.W.W. goal, Rothko would not have been an ideal Yale student. Moreover, he entered Yale in 1921 when America, still under the shock of the Russian revolution, was responding apprehensively. There had been the Palmer raids in 1919 in which many young immigrants coming from the same background as Rothko had been rounded up, humiliated, and sometimes deported. A permanent Red scare was initiated by Mitchell Palmer which spilled over into renewed assaults on organizers of labor. If dreams of music and the arts accompanied Rothko as he made his way through two years of Yale, probably other dreams—of social equality, for instance—were just as insistent. The ideal of the self-educated anarchist seemed to lie behind his decision to take off for New York, as he said, "to bum about and starve a bit."

In New York he took a room with a Mrs. Goreff at 19 West 102nd Street, and began his adult life. Surviving at odd jobs in the garment district, including that of a cutter, and, as was often the custom, taking a little help in the form of a bookkeeping job with a relative, Rothko at twenty seemed resourceful. No doubt the gospel of the anarchists fired his ambition to experience the life of a worker, but it wasn't long before he found his way to the Art Students' League. By January 1924, he was enrolled in George Bridgman's anatomy course, and probably took a sketch class as well. Still undecided, the twenty-one-year-old Rothko returned to Portland for a time where, as he often proudly recounted, he joined a theater company run by Josephine Dillon, wife of Clark Gable. The theater experience seemed important to Rothko although he apparently made no attempt to stay in the theater but returned to New York a few months later and was soon enrolled again at the Art Students' League.

This time, however, Rothko enrolled in the class of Max Weber, taking both Weber's still-life class and another in life sketching. Weber was then forty-four years old, a Russian-born Jew who had, like Rothko, come to America at the age of ten. He was considered a star at the League, where students flocked to his classes and respected him for being a "modern" artist. In fact he was a living repository of modern art history. He had ventured to Europe while still a student to seek out the work of El Greco, then considered a rare precursor of modern art. He had studied the masters of the early Renaissance in Italy, and in 1905 had settled in Paris just in time to see the scandalous salon in which Matisse, Vlaminck, and Derain struck their blow against Impressionism and called upon themselves the wrath of the critics, who dubbed them wild beasts. The twenty-four-year-old Weber was only too ready to accept the rebellious French avant-garde and enthusiastically responded to its interest in the art of primitives (even French so-called primitives— he became a friend of Rousseau). Above all, he shared their interest in Cézanne who had finally, after his death, come to be regarded as a genius and father of the modern spirit. Matisse also exercised influence on Weber, who was one of the first pupils in Matisse's short-lived school. "In 1908, when Weber returned to America," wrote James Thrall Soby, "he brought with him four dominant impressions of his years of study in Europe. He remembered above all the sinuous Mannerism of El Greco, the solid masonry of Cézanne, the bold, sweeping arabesques of the *fauves*, the searing distortions of tribal sculpture."[4] To this Weber added the newly arrived Cubist approach within a very short time, thus covering every aspect of the foundations for 20th-century art. By the time Rothko encountered him, Weber was already celebrated for his early works in a primitivistic manner, where echoes of tribal art were frequent, and for his adaptation of the Cubist spatial canon to an essentially Expressionist vision. He had left Cubism behind, however, and during the 1920s Weber concentrated on expressionist figure studies in landscape settings and still-lifes. As a figure painter, he emphasized the expressive gesture of the nude. Although the inspiration is definitely Cézanne, particularly in his rendition of mountains, the atmosphere of Weber's grouped nudes is tempered by a consciously expressionist goal. The nude was to be the vehicle of the artist's emotion. And its environment was to be an evocation of mood.

In teaching, Weber stressed the importance of the figure in the history of art, and above all, the importance of interpretation. He was however still a "modern" artist and shrank from the classical reaction that had overtaken both Europe and America after the First World War. The great "call to order," encouraging a return to naïve realism and to mimesis or "objectivity" as the basis of painting was antithetic to Weber's early modern idealism. Weber had, after all, been one of Stieglitz's band of intrepid modernists. He had fought the philistines side by side with Marin, Dove, and Hartley. Later, he had tried to infuse the Cubist idiom with expressive subject matter, often drawn from Jewish themes. A man small in stature, he had enormous ambitions, and strong opinions. The young Rothko was clearly in his thrall.

Rothko's earliest sketches show him striving to find expressive distortions of the figure, clumsily and hesitantly, with Weber's own work in mind. Weber's allusions to the short, stumpy female figures of certain African tribes, from which Picasso had drawn; and his memory of Cézanne's bathers, were assimilated by Rothko who was clearly straining to learn a language adequate to his as yet unrefined strong feelings. Interspersed with the studies of conventional nude poses in Rothko's earliest sketchbooks are a few drawings of mothers and children, showing the young Rothko tentatively expressing a tenderness in terms of subject. His still-life essays also showed con-

1. Untitled   no date, probably early 1930s

2. Untitled   no date, probably early 1930s

3. Untitled   c.1925-27

siderable awkwardness, even for a beginning art student. But it was clear that during the two years he spent sporadically studying with Weber, he had absorbed many principles from which he could move on in experiment. Weber's views were available in a small book published in 1916.[5] To judge by the work of his students and their reminiscences of his classroom talks, his basic ideas had not changed by the time Rothko attended his classes. He said such things as:

> Always it is expression before means. The intensity of the creative urge impels, chooses and invents the means. . . . But the wonder and blessing of the spirit is that it can manifest itself best through the simplest means and that it knows no technique. . . .

> The imagination or conception of an arrangement of forms or of a particular gamut of color in a given rectangle is not a matter of means, but an inner spiritual vision. . . .

> Expression is born in stillness and in solitude. It is as if one were listening to time and telling what he hears.

In his experiences at the League, Rothko for the first time had a taste of the nature of artistic life in New York. He made a close friend among his fellow students, Louis Harris, and probably at Weber's instigation, began to make the rounds of the galleries. The expressionist mood at the Art Students' League was informed by a few exhibitions from Europe presenting the work of Rouault as well as the German Expressionists. In 1923 J. B. Neumann had opened his "New Art Circle" on Fifty-seventh Street with his motto emblazoned over the door: "To Love Art Truly Means To Improve Life." Neumann had been a successful young dealer in Germany and now was preparing to introduce to America such rising stars as Paul Klee, Ernst Kirchner, and Max Beckmann. In addition, he was eager to find American artists and soon became Weber's dealer. Neumann really believed that art could improve life and worked tirelessly to persuade the often recalcitrant American public. He was a man of vast interests and boundless enthusiasm who, in his intermittent publication *The Art Lover*, "devoted to the neglected, the misprized, and the little known," published reproductions of such various works as African sculpture, ancient sculpture, northern Eu-

4. Untitled   c.1929

ropean Renaissance woodcuts, American Indian and colonial American works along with the latest works of Klee, Max Beckmann, and Weber, or an occasional El Greco. He was always cordial to students, often taking them into the back room to see his latest acquisitions, talking to them of his encounters with great minds, and referring them to books in which indispensable information or inspiration resided. He was an enthusiast.

After Rothko ceased to attend the Art Students' League regularly, he worked doggedly to improve his drawing, and to find adequate means to express his temperament. His sketchbooks show him trying to compose and invent as well as to record. He worked with the same motifs as many of his contemporaries ranging from landscape to still-life, from interiors to genre scenes. Evidence that he was familiar with Weber's small gouaches appears in the late 1920s: generalized figures in expressive postures rendered in a hazily romantic atmosphere. These early works were never literal. Even cows and horses were portrayed as taking part in some special mood and deliberately painted in a quasi-primitive mode. Rothko also tried his hand at depicting the urban environment, such as the Elevated, or scenes of life in the crowded quarters of the poor where most of the young artists lived.

His connection with the ASL was apparently maintained. In 1928 Bernard Karfiol, an instructor at the League, assembled a group show for the Opportunity Gallery in which he included both Rothko and his fellow student, Lou Harris, as well as Milton Avery, a forty-three-year-old painter who had come to New York three years before. Rothko had met the first requirement of the professional—exhibition—at the age of twenty-five.

2

Rothko always maintained that he was self-taught, and that he had learned in the beginning from other painters. However, Weber presided over his early development. Weber's stress on expression was the guiding principle. While many years later Rothko would repudiate self-expression as the proper function of art, during his formative years he drew upon that part of the modern tradition that stressed the reflection of individual personality. Through Weber he had learned to respect the direct expression of feelings evident in the work of the untutored and children. When he found a job in 1929 at the Center Academy in Brooklyn—a school for children run by a synagogue—he put his principles to work. Although friends sometimes point out that Rothko didn't like teaching, there is evidence that his work with children at the Center Academy was important to him. It enabled him to test his theory and reinforce his growing belief in art as man's expression of his total experience in the world. As he noted in a sketchbook of the mid-1930s, in which he was apparently preparing a speech to celebrate the tenth year of the Center Academy's art program, modern art had made the public aware of children's art by which they could learn "the difference between sheer skill and skill that is linked to spirit, expressiveness and personality." In a fragmentary sentence, he tried to be more specific: "Between the painter who paints well and the artist whose works breathe life and imagination." This rebellion against academic method reflected Rothko's earliest attitudes. In the older methods, Rothko noted, children were given examples and expected to perfect themselves in imitation of them. In his approach, "the result is a constant creative activity in which the child creates an entire child-like cosmology which expressed the infinitely varied and exciting world of a child's fancies and experience. . . ."

Rothko's serious study of the way children create and the great value he placed upon his teaching experience can be gauged by the careful notes he made during the more than twenty years he taught at Brooklyn's Center Academy. He conscientiously read the growing literature on art education and was familiar even with obscure sources. For instance, in an undated note he cited the pioneer Franz Cizek, a source that only a dedicated student of the psychology of art would have encountered.[6] Cizek had begun his experiments with children in Vienna in the late 19th century. In 1910 he organized classes for children from four to fourteen years in which they were free to work in many media and formats protected against "corrections" and "improvements." One of Cizek's maxims was "*Werden, wachsen, sich wollenden lassen*" (becoming, growing, achieving fulfillment). His work was presented at a congress in London in 1908 and thereafter gained currency in specialized English-speaking circles, and from 1920-22 an exhibition of works produced in his classes traveled in England and possibly in America. This might have come to Rothko's attention through subsequent discussions. In any case, his deep interest in what Cizek had called "the blooming and unfolding (Aufblühen un Entfalten) of the artist in the child" was certainly spurred by Rothko's need to understand his own impulse and to ponder the nature of art, as well as by his missionary character. Rothko's notes on child education are invariably fervent, filled with indignation at the insensitive general practises in art education, and generally reflect his need for a cause. Years later he would speak of abstract art in much the same way: as a cause.

It is not surprising that Rothko spoke of the "cosmology" of the child's world, or that he valued "spirit, expressiveness and personality." All through the 1920s theories of personality were discussed constantly. Freud's lectures in America in 1911 and his publication in English in 1915 of "On the Interpretation of Dreams" inspired voluminous commentary. The Jazz Age took to Freud with alacrity. In Bohemian circles people analyzed each other constantly, and warned each other of the dangers of repression. Malcolm Cowley stresses the importance of the idea of "self-expression" during the 1920s in *Exile's Return* published in 1934. The concept of self-expression, he summed up as: "Each man's, each woman's purpose in life is to express himself, to realize his full individuality through creative work and beautiful living in beautiful surroundings." As a corollary, Cowley added, "Every law, convention or rule of art that prevents self-expression of the full enjoyment of the moment should be shattered and abolished. . . ." Rothko's interest in self-expression was perhaps shaded by his instinctive interest in finding an image of a whole, a "cosmology," but in his youth, expressiveness and what he called personality, seemed the highest values. In referring scornfully to "sheer skill," Rothko placed himself among the artists in New York's vanguard who had come to despise the empty technical feats of the American academic ranks—they were legion in those days—and who had also questioned the value of the elegant postwar European variations on early modern idioms. The results of French cuisine, as they slightingly referred to the masses of well-painted but decorative French works, were not for them. Nor were the clumsy works of the so-called regionalists very appealing. For Rothko and his friends, self-expression meant largely self-expressionism, and they looked not only to the German Expressionists and Soutine and Chagall, but also Rouault, whose somber watercolors seemed to fulfill demands both for expressivity and for skill.

In his late twenties, Rothko was beginning to seek principles which would help him to define his task as an artist. He was still moving fitfully in many directions and with considerable confusion when, after the initial meeting at the Opportunity Gallery with Milton Avery, he was brought to Avery's studio, probably by the musician Louis Kaufman who was a friend from Portland. Rothko's first encounter with a committed professional artist had been with Weber, an intense, voluble, and often bitter little man whose opinions were always emphatic. Weber wished to be a maestro. Avery, on the other hand, was exceedingly quiet, modest in appearance and gesture, and exuded calm confidence. He was then in his forties and had been painting steadily for a number of years. He and his wife Sally, who was as exuberant as he was self-contained, lived modestly in a large one-room apartment where Avery got up every morning and painted, as he did in all the subsequent apartments the Averys maintained. The steady, unswerving devotion to his task that Avery showed Rothko was of inestimable value to the younger artist. He had been longing, from his early childhood, for a means to a total commitment. In Avery's way of living his life as a painter, Rothko found a mode. As he was to write after Avery's death, for the young who were questioning and "looking for an anchor," the admission to Avery's studio was to have "the feeling that one was in the presence of great events."[7] No doubt the cheerful family life of the Averys, whose home was always both a studio and a home, and the stability of the marriage appealed to Rothko. He was living then, according to Sally Avery, with Lou Harris on the Lower East Side in a tenement apartment with the toilet in the backyard. Crowded as it was, with a street life that never subsided, with cries of vendors, open markets, and hordes of children; with the din of Yiddish argument and eternal discussion, the Lower East Side contrasted sharply with the calm of Avery's studio. "I cannot tell you what it meant for us during those early years to be made welcome in those memorable studios on Broadway, 72nd Street and Columbus Avenue. . . . The instruction, the example, the nearness in the flesh of this marvelous man—all this was a significant fact—one which I shall never forget."

What Rothko perceived in Avery was his "naturalness" and his quiet poetry. No doubt he valued the naturalness, "that exactness and that inevitable completeness which can be achieved only by those gifted with magical means" because he himself was still so much a novice, and still trying so hard to find pictorial means to express his own poetry. He admired Avery's simple approach to the subjects of his paintings, drawn as they were from "his living room, Central Park, his wife Sally, his daughter March, the beaches and mountains where they summered; cows, fish heads, the flight of birds; his friends and whatever world strayed through his studio: a domestic, unheroic cast." Rothko could address himself only with the greatest difficulty to the modest and domestic. Much as he longed to, and much as he appreciated Avery's ability to live within the terms of the perceived world, there was, in him, a strong pull toward the cosmic and heroic. All the same, the salubrious environment of Avery's studio where he was always welcome, and Avery's own canvases, inspired him. Rothko diversified his work from the moment he began frequenting Avery.

Avery himself had not yet found the harmonious condensations that appeared in later work. His drawings tended toward a kind of illustrator's literalness, with awkward cross-hatching. His paintings, though, already evidenced his interest in expressing his subject through suggestive color harmonies and simplification of gesture. He foreshortened or elongated figures, sometimes stressing the thrust of a limb, or the tilt of a head by means of gentle distortion. For Rothko, Avery's generalizing of such irking details as fingers and feet would have been a blessing, as he had not had the long training in drawing of the average art student. Avery's reclining figures on the beach, or his child portraits, displayed a tenderness that Rothko valued highly. Rothko's watercolors and gouaches, once he met Avery, begin to show a clarification of purpose. He was working hard to fulfill an intense desire to establish a prevailing mood. Distortions similar to Avery's, such as making large bodies with small heads, or small heads on large bodies, appear around 1930. Increasingly Rothko strived to compose appropriate settings. His colors were often muted, with terra-cottas and whitened browns and grays as in certain Avery paintings of the early 1930s.

Around the same time Rothko began to see a lot of Adolph Gottlieb, a painter his own age who had had the good fortune to wander in Europe while a student. Gottlieb's temperament was less romantic, perhaps, but he shared with Rothko an intense interest in finding an expressive language of his own. Unlike Rothko, he did not think of himself as a born anarchist, although he was quick to challenge authority. They had met at the Art Center Gallery, one of the very few small galleries that introduced new artists, and later at the Opportunity Gallery where Rothko had first showed, and where each month a well-known artist presented new young artists. Gottlieb remembers Kuniyoshi and Alexander Brooke as selectors and that "Mark and I and Milton Avery frequently got into those

5. *The Bathers* c.1930

shows."[8] Gottlieb and Rothko's friend Lou Harris, began to make Avery's home an almost daily meeting place. Sally Avery remembers long evening conversations, and animated discussions about the paintings on the Averys' wall. Rothko was "very verbal and told fabulous stories, a continual raconteur."[9] Although Avery was more than eighteen years older, he readily allied himself with his two young admirers, and submitted to their criticisms of his work. This shared experience seemed to buttress Rothko's confidence. He and Gottlieb continued the pattern for years, even joining the Averys summers for vacations. They would paint furiously during the day and then compare works. The long evenings were spent in discussions, often of literature—poets such as T. S. Eliot and Wallace Stevens—and, according to Sally Avery, some philosophy. Rothko, she says, was reading Plato.

Both Rothko and Gottlieb had a great need to talk and would meet frequently. "Rothko was one of the few guys who were articulate," Gottlieb recalled, "because in those days painters were sort of silent men." No doubt they talked about the works they saw in the newly opened Museum of Modern Art, and they talked about the distasteful, chauvinistic art critics who fulminated against "foreign" influences in American art, stridently calling for an American art that the man in the street could identify with. The American Scene painters were very much on the scene. One of them, Thomas Hart Benton, converted from abstraction and took up the banner of Americanism. He attacked European sources and, as he wrote shortly after, at the height of the Depression, he separated American art "from the hothouse atmospheres of an imported, and, for our country, functionless aesthetics." He went on to speak of "releasing American art from its subservience to borrowed forms."[10] Neither Rothko nor Gottlieb could accept such rhetoric, although they, like many others, were watching social and political events in America with great attention and spent many hours discussing the national crisis initiated by the crash in 1929. During the early 1930s Rothko and Gottlieb were more aware of what they refused than satisfied with what they themselves undertook. They refused American regionalism, Cubist decoration, total abstraction, and the sharp realism of the New Objectivity. They were still moved by moody European Expressionism, but had been smitten with Avery's version of the rhapsodic art of Matisse. They exposed themselves to many kinds of painting, and occasionally the distinct impact of one or another modern (or sometimes not so modern, as in the case of El Greco elongations, or light effects of Corot) could be seen in their work. They read the prominent art journals of the day and scanned the reproductions. In *Creative Art* in 1931 and 1932, for instance, they could become acquainted with reproductions of recent work by Max Beckmann, Picasso, Derain, and the French Surrealists. At the time there were also articles about the Italian painters of the *Valori Plastici* group—those who had programmatically renounced abstract art—and the *scuola metafisica*. Reproductions of de Chirico appeared fairly often in art periodicals, and also of Carrà. The somewhat softer metaphysical paintings of Carrà may have attracted Rothko. In one of his gouaches dating back to the late 1920s or early 1930s, there is an enigmatic juxtaposition of a large almost triangular head with painted head just behind it, its features rendered much in the stylized Carrà manner. There are also still-lifes and street scenes in which the whitened tones of some of the Italians, among them, Morandi and de Pisis, appear. For artists in the late 1920s and early 1930s, the various postwar Italian idioms may have seemed an ideal compromise between the radical modernism of the non-objective painters and the stultifying naturalism of the painters of the American Scene School.

In 1932 Rothko married Edith Sachar, a young jewelry designer who managed to make a living even during those difficult days. They lived near the Averys although the new young wife was apparently resistant to Avery's influence. Rothko had his first one-man show the following summer at the Portland Museum, where he took the unusual step of showing his own work along with works by the children in his classes at the Center Academy. It was clear that he believed wholeheartedly in the values inherent in the work of the uncorrupted. Child art, for Rothko, was a touchstone, a kind of barometer of truth. In this he may well have been encouraged by J. B. Neumann, whose advice and approval Rothko valued. Several of the children's drawings in the exhibition were in the collection of Mrs. Neumann. Rothko himself showed watercolors and drawings.

The critic for the *Oregonian* wrote approvingly (July 30, 1933) of a strong Cézanne influence. Rothko had painted a group of watercolors with the rationalized planes and spareness associated with Cézanne, but, as Diane Waldman has pointed out, these may well have been youthful imitations of work he had seen by John Marin. More likely, they were paraphrases of Avery's system, itself derived from Cézanne. The following November, Rothko had his first one-man show in New York at the Contemporary Arts Gallery. The exhibition included oils, watercolors, and drawings, and must have spanned several years. A critic in *Contemporary Arts* discussed one painting, "The Nude," remarking on its "ponderous structure" that harks back to the "Eight Figures" by Max Weber. But the critic also noticed the Cézanne influence. There were probably other influences as well. Between 1930 and 1933 Rothko had been emboldened to try more complex compositions, sometimes with several figures. Matisse's voice, filtered through Avery, is faintly heard in some of the paintings referring to music. There is one painting of a group of figures listening to a pianist in which a bright blue rug is painted parallel to the picture plane, and a geometric element—a square—used to balance the composition. On the other hand, there is an oil on masonite painting of a string quartet in which somber blues and browns predominate and the heads are rendered in ghostly gray tones, indicating Rothko's express wish to convey his feelings about the music. In the street scenes, Rothko used architectural settings to enhance a vision of urban isolation and loneliness. The theatrical effect is apparent, with scenes brought close to, and parallel with, the picture plane and backgrounds seeming like theater flats. In one, two women and a child are centered in the foreground flanked by two long narrow niche-like windows in which elongated robed mannequins are indicated. Still bolder is a painting that must have arrived after one of his forays in the Metropolitan Museum. This small painting on masonite seems to be an interpretation of a Renaissance interior. It is composed in two tiers, with an ambiguous portrait, perhaps of a cardinal, dominating its center and two apparitional marble sculptures flanking its lower portal. Three darkly scumbled figures are in the doorway. A drama is suggested here, and Rothko depended on the sketchy ambiguities in his brushwork to

6. Untitled   c.1932

7. Untitled   c.1933

reinforce the dramatic theme. While the flatness reminds us of his modern preoccupations, the color—oxblood above, green and dark brown below—and atmosphere remind us that Rothko was an admirer of Rembrandt and Rouault.

While Rothko was struggling to find his voice, there were serious distractions. The entire artistic community was beginning to feel the strain of the Depression. In the neighborhoods that many painters inhabited in New York there were unmistakable signs of the débacle: breadlines, apple sellers, and soapbox orators. All America was discussing the unprecedented economic collapse. The atmosphere of emergency sponsored unusual attitudes and opened America to a frantic debate. Probably the crisis rekindled Rothko's political energies, although he had already decided that his first commitment was to art. He and his friends were aware not only of the domestic disaster but of sinister developments on a world scale. All during the 1930s there was plenty to worry about: the consolidation of Mussolini's power; the rise of the Nazis; Italy's conquest of Ethiopia; the 1936 outbreak of the Spanish Civil War, and Hitler's arming of the Rhineland; the Moscow trials in 1937; Hitler's annexation of Austria and his occupation of Czechoslovakia. On the personal level, there was imminent starvation, even for the resourceful Bohemians who found themselves, in 1934, without means. There were also grave threats to civil liberties. In 1934 an event occurred that roused the artists to action: Diego Rivera, much admired at the time, had completed his murals for Rockefeller Center. When the Rockefellers discovered that he had portrayed Lenin, they ordered the murals destroyed. Agents arrived on the site where Rivera was putting the finishing touches and ordered him off the scaffold. A few assistants present ran to telephones to alert other artists, and a large group including Rothko assembled to protest. They tried, unsuccessfully, to block the demolition of the work. Out of this dramatic night's gathering grew the Artists' Committee of Action. At the same time militant artists who foresaw even more dire consequences of the Depression had managed to organize the Artists' Union at a large meeting attended by Rothko and Gottlieb, among some two hundred others. The Artists' Union's purpose was "to unite artists in the struggle for economic security and to encourage wider distribution and understanding of art." There is every reason to believe that Rothko's old commitment to "one big union" was bestirred. He was present at the Union's monthly meetings, and when the publication *Art Front* began in November 1934 he was active among its supporters. Among the letters published in the first issue encouraging the editors was one from Lewis Mumford praising the fight for a municipal art gallery "because it will help take art out of the sphere of mere connoisseurship and wealthy patronage," and one from Max Weber, whose presence in Rothko's artistic life was still evident. Weber wrote, "My heart is full; welled to the brim with resentment for I see clearly—as other artists who are socially conscious—how Nazism, chauvinism and fascism are worming into the life of art and artists. . . ."

The Artists' Union was demanding government projects that would be more effective than those of the newly established Public Works Administration, with "complete freedom in the conception and execution of work." Rothko was active in demanding the municipal gallery, and in struggling with the forces in the Artists' Union that mistook provincialism for a new brand of American art. No doubt he experienced the chaotic political events intensely, and, like so many others, hesitated before their enormities. Both Juliette Hays, wife of the writer H. R. Hays, and Joseph Liss, a writer interested in theater, who had met Rothko around 1935, remember Rothko's intense concern with the political situation. There were long and heated discussions in which Rothko expressed his conflict about taking positions. As an artist he felt reluctant to join any groups, but as an individual concerned with social justice, he felt obliged to support group activities. Aesthetically he was clear: he loathed every thing that smacked of social realism; fulminated against such favored figures as Joe Jones and William Gropper, whom he regarded as little better than cartoonists. Although the long evenings of discussion often veered to politics, Rothko, "who liked to probe a situation looking at it in many ways" according to Liss, liked best to talk about painting. He was, in those days, open to all kinds of aesthetic experience, according to Mrs. Hays, and thought of himself as a vanguard artist. The elaborate conversations, she reports, wandered throughout art history, and Rothko's "fertile mind" was always at work. "He was ebullient. He was a real force." Mrs. Hays points out that in those days, Rothko had a great sense of humor and when the spirit took him, would parody the sentimental poetry of Edna Saint Vincent Millay by translating it into Yiddish. Liss, who like Rothko was Russian-born, remembered "kidding around" in Russian. In politics, he says, Rothko was avantgardish—more "an Art for Art's Sake type." Few artists could resist the need to make a statement reflecting the times, and Rothko, while he was apparently convinced about the separation of civic life and art, nonetheless adopted themes that were common to the epoch. His anarchist instincts protected him from succumbing to the vast surge toward social realism. Freedom of expression was after all the rallying cry of the old anarchists. But the need to reflect on the melancholy, even dangerous state of affairs was there, and crept into his work between 1934 and 1938.

As hectic as was the public life of most artists during the worst of the Depression there was paradoxically a suddenly rich artistic life as well. For Rothko there were to be many small exhibitions, and an increasing circle of artist friends sharing his attitudes. Moreover there were exhibitions that bolstered his innate distaste for the new social realism and regionalism. Julien Levy had opened his gallery, bringing in exhibitions of the Surrealists, first a group show and then a one-man show of Dalí to which Rothko responded enthusiastically. The second issue of *Art Front*, in January 1935, included a review by the erstwhile Russian painter John Graham warmly commending the gallery and saying of Surrealism that

> it is, as all abstract art, truly revolutionary, since it teaches the unconscious mind—by means of transposition—revolutionary methods, thus providing the conscious mind itself with material and necessity for arriving at revolutionary conclusions. . . . Those who think abstract art is just a passing phase are mistaken. Rivers do not flow backward.

In the same issue Stuart Davis, extremely active in the Artists' Union, and regarded as a left-wing militant by most other artists, reviewed the Dalí exhibition with astute appreciation. Subsequent issues contained lively debates between avowed Marxists, eager to maintain social content in visual art, and independent artists attacking the American scene painters and narrow visions in general. The ferment was rapid. Numerous arguments punctuated artistic life. The struggle to dominate *Art Front* saw almost monthly upheavals with which Rothko was thoroughly au courant. Early in 1934, he had become friendly with Joseph Solman who was to lead a protest a few months later initiated by several progressive artists, among them Rothko, against what they felt to be the narrowly economic and social interest of the publication. Rothko's work appeared in several group exhibitions, among them one at the newly opened Gallery Secession (an overt allusion to the expressionist bent of its exhibition policy). Among the artists in the first exhibition, in December 1934, were Ben-Zion, Ilya Bolotowsky, Louis Harris, Rothko, Gottlieb, Solman, and Nahum Tschacbasov. The convergence of these artists, several of whom were Russian born and most Jewish, indicates that Rothko still moved in the circles where he felt comfortable and familiar, and that his aesthetic remained largely within the principles set out by Weber, although he was already beginning to ponder other questions. Ben-Zion, for instance, was deeply committed to the Jewish experience and painted tableaux from the Old Testament, especially solemn patriarchs whose symbolized gestures were stressed in skillful exaggeration of hands. (Rothko perhaps was tempted by these hands, for in several of his subsequent semi-abstract paintings, the hieratic gestures of Ben-Zion's prophets appear with similarly drawn hands.) Tschacbasov also dealt with Jewish themes, and had been visibly influenced both by early Chagall and by the Burliuk brothers' primitivism.

"Somewhere around 1935," Gottlieb recalled, "we formed The Ten. We were outcasts—roughly expressionist painters. We were not acceptable to most dealers and collectors. We banded together for the purpose of mutual support."[11] The initial group came together in the Gallery Secession exhibition and then decided to form its own exhibiting society, hoping to persuade different galleries to hold their shows. They managed to show frequently for five years. As a group they fluctuated, although Rothko, Gottlieb, Harris, and Solman were constant. In recalling the years of The Ten several of its members have noted the importance of the monthly meetings in each other's homes, where discussions grew heated and lasted

33

nrough the night. Rothko was one of the most enthusiastic participants, and loved to follow labyrinthine paths in argument. Subjects varied, but certainly the strangely heightened years of the 1930s brought extremely varied issues to the fore. A great deal of the discussion was devoted to the perplexing artistic issues emerging within the troubled world context. Rothko seemed clear about his commitment to modern art, and also to certain artists of the past (Solman remembers Rembrandt, Bellini, and Corot). The entire group, who regarded themselves as "progressives," were keen followers of the news from Europe—the latest works of Picasso, Matisse, Rouault, and the German Expressionists. They paid close attention to the exciting exhibitions at the Museum of Modern Art, particularly the *Surrealism and Dada* show in 1936. They maneuvered among conflicting directions, trying to find a middle ground between the extremes of social realism and abstraction. "The whole problem seemed to be how to get out of these traps—Picasso and Surrealism—and how to stay clear of American provincialism, Regionalism and Social Realism," according to Gottlieb. For a time, *Art Front* stimulated wide-ranging discussions, having been rebuked by Solman—who represented, among others, Rothko, Bolotowsky, Balcomb Greene, and George McNeil—with a manifesto complaining that the editors were unaware of the educational value of the Museum of Modern Art, and that the magazine should look like an art magazine and not only a union newsheet.[12] Solman was soon invited to the editorial board and initiated a lively program. One of his first acts was to publish in the December 1935 issue a lecture given by Fernand Léger at the Museum of Modern Art. Almost certainly this lecture had been one of the hot topics at the monthly meetings of The Ten. Léger began by summing up the past fifty years as a struggle of artists to free themselves from old bonds, particularly the restraint of subject matter. The Impressionists, he said, had freed color, and "we have carried the attempt forward and freed form and design." He declared that subject matter was at last "done for" and stated categorically "that color has a reality in itself, a life of its own; that a geometric form has also a reality in itself, independent and plastic." Perhaps Rothko pondered Léger's central statement that the ridiculous question "What does that represent" was no longer relevant:

> There was never any question in plastic art, in poetry, in music of representing anything. It is a matter of making something beautiful, moving or dramatic—this is by no means the same thing.

This lecture was translated by Harold Rosenberg, whose own anarchistic instincts led him into battle with the *Art Front* executive board. They soon pushed him out.

The Ten meanwhile arranged their first group show at the Montross Gallery in December-January 1935-36, and a few weeks later they were part of the opening exhibition at the much fought-for—and won—Municipal Art Gallery, where all of the participants had demonstrated noisily for a revocation of the Gallery's Alien Clause, which stipulated that only citizens could exhibit there. A few months later The Ten had an unprecedented stroke of luck: they were invited by Joseph Brummer to exhibit in France. For "progressives," this was an exceptionally exciting opportunity to repudiate the rising nationalistic clamor in the ranks of the artists (some of whom were at work, along with Rothko, in the easel division of the WPA). The exhibition at the Galerie Bonaparte in November 1936 was introduced by a well-known Paris critic, Waldemar George, who noted that Rothko, in his "Crucifixion," revealed his nostalgia for masters of the Italian Trecento. A critic reviewing the exhibition felt that all the members had been influenced by Picasso and Rouault, and that Rothko's paintings "display an authentic coloristic value."

The attention paid to their public exhibiting life did not inhibit The Ten from performing volubly at Artists' Union meetings. They were active in preparation for the American Artists' Congress announced for 1936 by Stuart Davis in *Art Front*, December 1935, who gave as the objectives

> to point to the threat of the destruction of culture by fascism and war; to point out specific manifestations of this threat in this country; to show the actual accomplishment of such destruction in those countries where fascism holds power; to show the historical reasons for fascism, and to clarify by discussion what the artists must do to combat those threats.

But, Davis insisted,

The objectives of the Congress are not only defensive. A prominent part in this discussion will be given to an analysis of contemporary art directions; to the historical role of the artist in society; to the relation of subject and form in art to environment. . . .

The Ten could readily support the Congress's goals and were enthusiastically present when the three-day session opened in February 1936. Rothko, who was sensitive to world developments from his early youth and anxious at all times about 20th-century excesses, certainly had caught the prophetic overtones of Lewis Mumford's opening address in which Mumford stated:

The forces that are bringing on war, that are preparing for larger and better economic depressions in the future, are at odds with the forces of human culture. The time has come for the people who love life and culture to form a united front against them, to be ready to protect, and gauge, and if necessary, fight for the human heritage which we, as artists, embody.

3

When the easel division of the WPA was fully under way, scores of artists in New York lined up for their monthly paychecks, or checked into headquarters, and were drawn into conversation with their colleagues. The isolation of the past, in which so many artists had languished or fallen by the wayside, gave way to a new and turbulent situation. Artists of various persuasions were thrown together and forced, by circumstance, to wrestle with urgent questions, both social and aesthetic. The Ten broadened their associations. Still, they remained a unit that represented a distinct point of view. Decades later Willem de Kooning, also on the easel project, remembered Solman and Rothko, and their vigorous claims for Expressionism. With so much hectic discussion, and much foregathering, it was inevitable that thoughtful artists undertook to sort out their views. Rothko, who was on the project for the year 1936, during which time the Museum of Modern Art staged the "Fantastic Art, Dada, Surrealism" exhibition, had not moved significantly in his painting, but was beginning to draw together the threads of his diffuse thoughts on the meaning of art. He still maintained, as he later observed about children's art in his mid thirties sketchbook, that mere skill was meaningless unless it served to express spirit and personality. But he was beginning to reflect with more assurance on the various theories put forward by modernists. Léger's statement—that modern art had gotten rid of subject matter—was emphatically rejected. Years later, when Rothko was already an abstract artist, he still resisted non-objective theory, vehemently stating, "I would sooner confer anthropomorphic attributes upon a stone than dehumanize the slightest possibility of consciousness."[13] The Expressionists, who regarded themselves as legitimately ensconced in the radical modern aesthetic tradition, also had to face the more adamant claims of the newly formed group, the American Abstract Artists, who organized in the fall of 1936, modeling themselves on comparable groups in Europe: the English "Circle" group and the French "Cercle et Carré." These artists, a number of whom were on the WPA, proposed to unite American "abstract" artists in order to call attention to "this direction in painting and sculpture." In their general prospectus published in 1937 they were careful to state, "we place a liberal interpretation upon the word 'abstract'," thus allowing such varied artists as Ibram Lassaw, Byron Browne, George McNeil, George L. K. Morris, Ilya Bolotowsky (still showing with The Ten), Alice Trumbull Mason, Giorgio Cavallon, and Josef Albers to participate. All the same, as Lee Krasner remembers, "they had a little purist academy of their own."[14] The strong thrust of those who had received and accepted the non-objective tradition of Mondrian soon brought them into dominance in the organization. The Ten, with their conception of an expressive art epitomizing life, could pit themselves against such theories with fervor, although the prevalence of social realism seemed far more despicable to them. Rothko probably wrote or at least formulated the statement for The Ten's exhibition in protest to the Whitney Museum's policy in November 1938. The exhibition, called *Whitney Dissenters*, would fight "a new academy" that

> is playing the old comedy of attempting to create something by naming it. Apparently the effort enjoys a certain popular success, since the public is beginning to recognize an "American" art that is determined by non-aesthetic standards—geographical, ethereal, moral or narrative.

Rothko characterized his group as

> homogeneous in their consistent opposition to conservatism; in their capacity to see objects and events as though for the first

time, free from the accretions of habit and divorced from the conventions of 1,000 years of painting.

The clue to Rothko's preoccupations lies in the phrase "their capacity to see objects and events as though for the first time." Although his work, and that of his confreres in The Ten was still attached to sometimes sentimental themes—the *Sehnsucht* of the romantic Expressionists—he was already pondering the lessons of the Surrealists. In February 1937 the painter Jacob Kainen, who wrote criticism for Art Front and was friendly with the group, had praised an exhibition of The Ten, noting that they attempted to reduce the interpretation of nature or life in general to the rawest emotional elements; had a complete and utter dependence on pigment; and showed an intensity of vision. Of Gottlieb and Rothko, he had praise for their "usual sober plasticity, keeping everything simply emotional." Significantly, Kainen linked Expressionism with the mounting fear of war, warning "we are closer to chaos than we think." Barely a year later, Rothko was writing of seeing everything for the first time, free from the accretions of habit. Like Kainen, he was sensitive to the world situation and felt the chaos descending. His work of the late 1930s reflects his preoccupations: an increasing need to find the pictorial language to express his intimations of disaster; his increasing preoccupation with the life cycle, and above all, his growing awareness of the tragic, in which death is an acknowledged actor.

Nearly all the "progressives" among New York's painters had been startled and unsettled by the Surrealist exhibitions at Julien Levy's gallery, and the Museum of Modern Art. They responded according to their temperaments. Rothko seemed most affected by the Surrealist displacement of time; its long prospects in which the life of the spirit, or mind, could reach back to millennial visions; to origins in which everything was new, fresh, authentic. The Surrealist reverence for myth, in which the commentary on existence is for all time, and presents simultaneities of time, satisfied Rothko's craving for release from the quotidian; his instinctive drive to free himself from the shibboleths of contemporaneity and to be mobile, as he said his children students were, in the realm of the imagination where there could be "an entire cosmology." In his middle thirties, Rothko was ready to shape his visions of forlornness, and the tragic, on grand and sober lines. The resurrection of myth during the late 1930s, experienced by many artists, was to have special consequences for him.

The Surrealists were not the only ones to appropriate ancient myths as a new framework for artistic discourse. The plasticity of the word "myth" allowed for considerable latitude in interpretation. The Marxists among the artists, for instance, had a fondness for the word. Marxist literary critic A. L. Lloyd, for instance, woefully twists and turns with the notion of myth, in "Modern Art and Modern Society" published in *Art Front* in October 1937:

> The bourgeois artist finds no theme he can acknowledge as valid. . . . This curious circumstance is largely due to the historic decay of myth, and it is the decay which has so materially contributed towards widening the gap between economic and spiritual production. . . . As Marx pointed out, the role of myth is to express the forces of nature in the imagination. . . . Only an autochthonous mythology, a mythology coming from the same soil, the same people, the same cultural superstructure of the same economic order can be an effective intermediary between art and material production, despite the bourgeois superstition that any mythology, even one personally transformed into a kind of private religion, can form such a link. . . .

To this, such self-styled anarchists as Rothko and Barnett Newman (whom Rothko began to see frequently during the late 1930s) responded with derision. For them, the realm of the mythic knew no "autochthonous" bounds. Other sources for the mythic approach to human history were more congenial. There was considerable discussion of the works of T. S. Eliot during the 1930s. Eliot was one of the authors Sally Avery recalls as having been read and discussed during the summer evenings when the Rothkos, Averys, and Gottliebs vacationed. Eliot was unpopular with the Marxists, who regarded his disdain for historical materialism as reactionary. But the artists in Rothko's circle were strongly attracted to Eliot's anti-historical critique implicit in "The Waste Land." It is quite possible that Rothko's enthusiasm for Frazer's *The Golden Bough* derived from his close reading of Eliot, who referred his readers in his notes to the anthropological classic. Eliot was aware of the needs from which his use of Frazer derived, as can be seen by his comment on James Joyce, whose *Ulysses* had appeared in 1922, the same year as

"The Waste Land." "In using the myth, in manipulating a continuous parallel between contemporaneity and antiquity, Mr. Joyce is pursuing a method which others must pursue after him. . . . It is simply a way of controlling, of ordering, of giving a shape and a significance of the immense panorama of futility and anarchy which is contemporary history."[15] Eliot's estimation of Frazer's importance was again stated in *Vanity Fair*, in February 1924, when he said that Frazer "has extended the consciousness of the human mind into as dark and backward an abysm of time as has yet been explored." Both *The Dial* and *Vanity Fair* were read by artists in Rothko's circle. What Rothko needed was not the literary allusions so much as a new way, as Eliot said, of controlling and ordering his experience in a time he found out of joint. For him, the atmosphere of removal was essential in order to formulate his point of view of existence. The myth for Rothko was, and remained, the source of a dramatic confrontation between nature, ruled by law, and the human imagination, free in its expression. His appropriation of myth sometime in the late 1930s derived from a need identical with Eliot's and Joyce's. He found in myth what Ernst Cassirer described as "a dramatic world of actions, of forces, of conflicting powers."[16] In every phenomenon of nature, Cassirer writes, the mythic world

> sees the collision of these powers. Mythical perception is always impregnated with these emotional qualities. Whatever is seen or felt is surrounded by a special atmosphere—an atmosphere of joy or grief, of anguish, of excitement, of exultation or depression.

Interestingly, Cassirer looked to the American philosopher John Dewey, for support of his argument. Dewey published *Experience and Nature* in 1925, and *Art as Experience* in 1934. Both books were widely discussed among American artists—far more than is generally conceded. Cassirer, citing the 1925 book, maintains that Dewey was the first to recognize and to emphasize the relative right of those feeling-qualities which prove their full power in mythical perception. He quotes Dewey:

> Empirically, things are poignant, tragic, beautiful, humourous, settled, disturbed, comfortable, annoying, barren, harsh, consoling, splendid, fearful; as such immediately and in their own right and behalf.

41

In *Art as Experience,* as William Seitz pointed out after extensive conversations with Rothko and other Abstract Expressionists, Dewey maintained:

> If the artist does not perfect a new vision in his process of doing, he acts mechanically and repeats some old model fixed like a blueprint in his mind . . . the real work of an artist is to build up an experience that is coherent in perception while moving with constant change in its development.

What Cassirer calls "feeling-qualities" became more and more important to Rothko, whose emotional life had thrust him from subject to subject and technique to technique without ever providing satisfaction. Painting was his only means, and it always fell short. Often he wistfully compared painting to poetry or to music, and with an uneasy sense of his own inadequacy. It was not only the "poignant"—a favorite word—that he sought to express. He also harbored a deep need to discover in the world, and therefore in his "self," the heroic—a quality of which the modern world seemed bereft; a quality perhaps impossible to achieve ever again. From the late 1930s through the 1940s, Rothko was preoccupied with the personal archaeology that would reach deep into what Eliot called the abysm. As he wandered toward his "self" he instinctively returned to what his complex temperament could thrive on. At once tender and impatient, profoundly in need of a framework that could contain his passion, he cast himself back to an era of sober grandeur. He read the ancient Greeks, above all, Aeschylus. The importance of Aeschylus in his psychological evolution is indisputable. In countless conversations, once in public forum, and in writing Rothko invoked Aeschylus (along with Shakespeare sometimes). Several of the important paintings of his so-called Surrealist period were based on the Agamemnon trilogy.

There are many reasons why, at a certain moment, a condition of acute sensibility and susceptibility to certain works of art arrives. These reasons are suffused throughout an artist's inner life and can only be faintly determined. When Rothko returned to the Greek tragedies, the circumstances, the state of the world to which he was never insensible, were as menacing as in ancient times. The blood that had flowed unstanched during the Spanish Civil War presaged the Hitlerian Holocaust. Artists, or at least certain artists, had dreaded a cataclysmic future for more than a decade already. Rothko's early efforts had been to convey largely melancholy responses to the human situation. Yet, the need for a loftier, somewhat more remote inspection was always there. Perhaps his early training, with its emphasis on the Law and Jewish fate, and its exposure to the stately poetry of the Old Testament, readied him for his profoundly moving encounter, sometime in the very late 1930s, with King Agamemnon. How was any sensitive artist to comprehend the dimensions of destruction encompassing the Western world in those days? Aeschylus, whose meditation on war and death in the Agamemnon trilogy was measured but passionate and whose understanding of human frailty was sutble, had found a form, or rather, a structure. The containment that so much enhanced the drama, was probably what Rothko recognized, finally, as a possible answer to his own artistic dilemma. Many years later Rothko would speak of "a clear preoccupation with death—all art deals with intimations of mortality." In Aeschylus, death takes its place as a *persona*, acted upon and acting, in a total drama.

Rothko was not a literary critic or an anthropologist or a psychoanalyst or a classicist. He read his Aeschylus not from a need to know but from a need to be moved. The "poetic"—which is to say the diction that Aeschylus and no other chose—is what moored itself in Rothko, enabling him to draw upon it as he undertook to work out his own destiny as a creator. Early in the drama, as the chorus files out before the palace of King Agamemnon to recite the woes that had overtaken the house of Atreidae ten years before, the very first image strikes to the heart—an image Rothko was to use in many works from 1939 to 1947:

> Their cry of war went shrill from the heart,
> as eagles stricken in agony
> for young perished, high from the nest
> eddy and circle
> to bend and sweep of the wings' stroke,
> lost far below
> the fledgings, the nest, and the tendance.

Translated by Richmond Lattimore

The pathos of the mighty eagle insulted is likened by Aeschylus in the conclusion of the same speech to the fateful condition of man who, "leaf withered with age / goes three footed / no stronger than a child is, / a dream that falters in daylight." Aeschylus' guarded critique of the gods addressed itself to Rothko who had early learned the significance of suffering. In declaring his Russianness, he betrayed his attachment to the Aeschylean notion that Zeus, who guided men to think, "has laid it down that wisdom comes alone through suffering" and that, as the chorus intones,

> From the gods who sit in grandeur
> grace comes somehow violent.

Not so different from the portentous passages in the Old Testament is the chorus's heart-rending reminiscence of Agamemnon's murder of Iphegenia, "lovely as in a painted scene, and striving to speak," and its conclusion that

> Justice so moves that those only learn
> who suffer; and the future
> you shall know when it has come; before then, forget it.
> It is grief too soon given.

Rothko, harking back to the violence from which he had shrunk as a child, and which he abhorred as a man, wished to find the poetic equivalent of his sentiments on the eve of the Second World War. Aeschylus more than any other knew the perfidy of "the god of war, money changer of dead bodies" who sent young soldiers back, "packing smooth the urns with ashes that once were men" and who expected the bereaved to praise him through their tears. But, Aeschylus knew, the people "mutter in secrecy, and the slow anger creeps below their grief." The undertones in Aeschylus are born in images, very often the image of the bird, with which the chorus began. Toward the end of the tragedy the chorus asks

> Why must this persistent fear
> beat its wings so ceaselessly
> and so close against my mantic heart?

And Cassandra, on the brink of her fate, laments

> Oh for the nightingale's pure song and a fate like hers.
> With fashion of beating wings the gods clothed her about
> and a sweet life gave her and without lamentation.

As she goes to her death, Cassandra addresses the chorus:

> Ah friends,
> truly this is no wild bird fluttering at a bush,
> nor vain my speech.

Finally, Clytemnestra's cry, "Oh, it is sweet to escape from all necessity!" lingered in Rothko's imagination, together with the Aeschylean warning that such a situation is against nature.

Resounding tragedy in the high poetic tone of the Old Testament, whose stately rhythms were early instilled in Rothko, aroused him. Just as Mozart, in *Don Giovanni* and *The Magic Flute*, to which Rothko always thrilled, could sound grand, tragic notes with an august finality while yet preserving delicate nuances, the small touches and images in Aeschylus that lighten but do not banish the underlying tragedy stirred him. If birds' wings flutter either wildly toward freedom or in a panic of fateful recognition, they are the birds of Greek and Old Testament configuration—images of the soul; stand-ins for the invisible, unpredictable resources of men; embodiments of dreams of freedom, and dreams of wholeness, as, in their roundness, as Gaston Bachelard poetically insisted, they are cosmos. These birds are not exactly the symbols, with their violent connotations, fetched up by the Surrealists from Freudian depths. Not, at any rate, for Rothko who eventually abjured their visible shape and reverted to ancient images, or "feeling-qualities," that stood for cosmos.

Rothko arrived at a semblance of assurance in wielding myth and was able to indicate clearly the way he thought of it in his statement accompanying the reproduction of "The Omen of the Eagle," 1942:[17]

> The theme here is derived from the Agamemnon Trilogy of Aeschylus. The picture deals not with the particular anecdote, but rather with the Spirit of Myth, which is generic to all myths of all times. It involves a pantheism in which man, bird, beast and tree—the known as well as the knowable—merge into a single tragic idea.

His way to the "single tragic idea," which was to remain his most persistent quest, was neither direct nor easy. At first, around 1938, he seemed to seek to transform the visible and present into the "Spirit of Myth" which he felt to be direct and unmediated. What

8. *The Omen of the Eagle*   1942

9. *Subway (Subterranean Fantasy)*   c.1938

10. Untitled   c.1936

11. *Subway Scene*   1938

he admired in Avery—the poetic directness—was still elusive in his own work. Tragedy and human fate certainly underlay his approach to the subway scenes in the late 1930s (a common enough motif among urban realists) in which he adopted various stylistic simplifications in order to stress the broader drama of individual isolation. Perhaps his Russianness—what he thought of as his Russianness: his melancholy and his moodiness—invaded his soul as he grappled with the problem of individual destiny. Perhaps he remembered Dostoyevsky's warnings in so many of his stories and novels that utter isolation leads to madness and tragedy. Yet, how could an artist, an underground man, circumvent it? In any case, Rothko's series of subway scenes were fraught with foreboding, and some have seen the beginning of his mythic explorations in them. Here are the beginnings of an Orphic descent. In what is probably the earliest scene, the figures are dwarfed by a cavernous no-man's-land, grouped as they might be on a stage, and dimly picked out with a memory of Rembrandt and Rouault, in what one critic called the "dirty expressionist" style. In two other versions, dated between 1936 and 1938, Rothko strives to impose a distanced vision of souls isolated in a silent netherworld punctuated with subway columns that are like scenery flats. Figures are elongated, generalized, and merge with the tectonic details in shallow relief as they might in ancient Greek stylizations. By 1938, Rothko had decisively abandoned expressionist ambiguities in favor of the clear, lighter-toned vision of descending levels called "Subway Scene." This painting is often singled out justifiably as a turning point for Rothko. He himself valued it, and in later years, when visitors asked to see his early work, it was one of the few canvases he ever brought out of the racks. He kept it by him for many years.

From subterranean allusion to freedom of fantasy was then a short step, which Rothko took around 1938. In order to make such a shift, he had to prepare himself intellectually. It was probably then that of all the approaches to myth being discussed (Freudian, Jungian, Marxist, Frazerian etc.) he returned to Nietzsche, hero of the anarchist vanguard in the 1920s.

4

By the 1930s Nietzsche's reputation had suffered many indignities, not the least being his appropriation and misrepresentation by Nazi academics. Only the most intrepid libertarians in America could bear to face squarely the implications of Nietzsche's writings known at the time only in awkward English translations dating back to 1911. Among artists, however, Nietzsche's reputation was secure, partly because the entire early modern movement had been excited by his emphatic address to the darker region of emotions hidden away in the psyche. Numerous late 19th-century philosophers had begun to depend on the observations of the newly established science, psychology, but it was Nietzsche with his extravagant and artistically heightened exposition of psychological phenomena who alerted the burgeoning modernists. His courage in examining the irrational realm did not fail to incite the poets, painters, and composers who, in the late 19th century, felt themselves ready to repudiate the rational tradition in the arts.

In the United States, Nietzsche's presence among artists had a substantial history, going back to Stieglitz's pioneering magazine *Camera Work*, and the Bohemians of the 1920s. For Rothko, it had possibly begun with his interest in anarchism. The old anarchists, especially Emma Goldman, had stressed at least one aspect of Nietzsche's writings: his insistence on the freedom from superstition. Since, as Nietzsche had dramatically proclaimed, God was dead. . . . The Surrealists had also shown obeisance to Nietzsche. There were increasingly frequent references to him in both French and American Surrealist publications from the mid 1930s to the early 1940s.

There were references to many other dissident voices of the proto-modern period as well, all attacking the positivist tradition of the early 19th century. Amidst this clamor, artists looked to many sources to confirm their own intuitions. Rothko's situation, however, was different. What the renewed interest in myth meant to him found its definition not in the broad scanning that the European Surrealists specialized in, but in an immersion in just one of Nietzsche's books—his very first, *The Birth of Tragedy*. It may well be that it was not so much Nietzsche's defiant re-casting of the definitions of ancient tragedy that attracted Rothko as the less emphasized emotional basis given by Nietzsche in the original edition's title: *The Birth of Tragedy Out of the Spirit of Music*. The young Nietzsche, like the young Rothko, had found immense emotional stimulation in listening to music. In Nietzsche's first book, he artfully sought to account for his own ecstatic emotions in an orderly, philosophic way, but the surging emotion derived from music rose up constantly, often in a poetic diction that presaged Nietzsche's later magnificent style. Rothko, who throughout his life not only listened to music but depended on it for solace, inspiration, release, was peculiarly attuned to the Nietzschean vision of the importance of music. While other painters were often interested in music, worked with music in the background, or even used musical analogies when they discussed their work, for Rothko it was so intimate a need that friends, who rarely agree on Rothko's basic characteristics, always concede that his response to music was fundamental. Not only was it always a presence in his life, but even when he visited others, he spent hours listening rapt, not even turning the leaves of a book. Herbert Ferber describes him lying on the grass in Vermont listening with total attention to the whole of *Don Giovanni*, while Carlo Battaglia remembers Rothko stretched out on a leather couch in Rome, windows open wide above the Renaissance plaza, wakefully attending the same Mozart opera. For Rothko, whose sensibility was so much excited by music, Nietzsche's text helped to focus disparate intuitions. From music to myth—as Nietzsche's flow of thoughts went in *The Birth of Tragedy*—is the natural trajectory of Rothko's own thoughts.

Nietzsche's conception of the aesthetic meaning of music was shaped by his study of Schopenhauer. Although he argued with the older philosopher on many basic propositions, he remained faithful to Schopenhauer's insights when it came to music, and quoted him at length. Some of these phrases would find echoes in Rothko's later attempts to clarify his vision in words. Certainly the general description of the nature of music in Schopenhauer can be likened to the radical decision in what Rothko later called his "enterprise." Nietzsche's choice of a passage from Schopenhauer's *The World as Will and Representation* indicated his need to be as specific as possible about the effects of music on himself. He quotes Schopenhauer as saying that music, if regarded as an expression of the world, is in the highest degree a universal language. Its universality, however,

> is by no means that empty universality of abstraction, but of quite a different kind, and is united with thorough and distinct definiteness. In this respect it resembles geometrical figures and numbers, which are the universal forms of all possible objects of experience and applicable to them all *a priori*, and yet are not abstract but perceptible and thoroughly determinate. All possible efforts, excitements, and manifestations of will, all that goes on in the heart of man and that reason includes in the wide, negative concept of feeling, may be expressed by the infinite number of possible melodies, but always in the universal, in the mere form, without the material, always according to the thing-in-itself, not the phenomenon, the inmost soul, as it were, of the phenomenon without the body. This deep relation which music has to the true nature of all things also explains the fact that suitable music played to any scene, action, event, or surrounding seems to disclose to us its most secret meaning. . . . When the composer has been able to express in the universal language of music the stirrings of will which constitute the heart of an event, then the melody of the song, the music of an opera is expressive. But the analogy discovered by the composer between the two must have proceeded from the direct knowledge of the nature of the world unknown to his reason, and must not be an imitation produced with conscious intention by means of

concepts, otherwise the music does not express the inner nature, the will itself, but merely gives an inadequate imitation of its phenomenon. . . .[18]

The "direct knowledge of the nature of the world unknown to his reason" was for Nietzsche, and later for Rothko, a most ardently desired knowledge, and Nietzsche went to great lengths to probe its meaning. In examining the existence of such direct feelings, Nietzsche posited his celebrated definition of ancient tragedy as a fusion of the Dionysian—the spirit of music, with its direct knowledge of creation's sources—with the Apollonian, the daylight revealing "the beautiful illusion of dream worlds in the creation of which every man is truly an artist." Nietzsche knew what Freud and Jung would later assert: that we dream in images.

> Thus the aesthetically sensitive man stands in the same relation to the reality of dreams as the philosopher does to the reality of existence; he is a close and willing observer, for these images afford him an interpretation of life, and by reflecting on these processes he trains himself for life.

The Dionysian, Nietzsche insisted (probably because he *felt*, rather than *thought*, in the presence of music), restored man to nature and "he feels himself a god . . . He is no longer an artist, he has become a work of art." He urges his readers to transform Beethoven's "Hymn to Joy" into a painting in order to understand what he means when he speaks of the Dionysian "feeling of oneness." Art, Nietzsche in his youthful enthusiasm thought, was a "complement and consummation of existence, seducing one to a continuation of life." But, like another very young *exalté* writing in the same period, Rimbaud, Nietzsche felt that there was a primal universe speaking through the artist. Rimbaud said, "*on me pense,*" and Nietzsche said,

> Insofar as the subject is the artist, however, he has already been released from his individual will, and has become, as it were, the medium through which the one truly existent subject celebrates his release in appearance . . . Only insofar as the genius in the act of artistic creation coalesces with this primordial artist of the world, does he know anything of the eternal essence of art; for in this state he is, in a marvelous manner, like the weird image

of the fairy tale which can turn its eyes at will and behold itself; he is at once subject and object, at once poet, actor, and spectator.

When Nietzsche came to criticize his own early work, he referred to the *Birth of Tragedy* as "rhapsodic" and recognized that it was "'music' for those dedicated to music, those who were closely related to begin with on the basis of common and rare aesthetic experiences"—a book for close relatives, as he put it. Rothko, by the nature of his inmost sensibilities, was a close relative. Never very firm in his images when they were wed to the immediate and concrete presence, he could become emphatic once he released himself to what Nietzsche, in his "Attempt at Self-Criticism" prefacing the 1886 edition, called a *strange* voice, a "spirit with strange, still nameless needs, a memory bursting with questions, experiences, concealed things after which the name of Dionysus was added as one more question mark."

> What spoke here—as was admitted, not without suspicion—was something like a mystical, almost maenadic soul that stammered with difficulty, a feat of the will, as in a strange tongue, almost undecided whether it should communicate or conceal itself . . .

It was in this strange tongue that Nietzsche uttered his conception of myth—an image that Rothko found totally congenial, and to which he adapted his own. Moved by the Aeschylean voice—the voice that Nietzsche said created a stage as "the mirror in which only grand and bold traits were represented"—Rothko accepted Nietzsche's version of the mythic in preference to others. He did not have to go to the writings of André Breton to find the fabulous, the miraculous, inhering in the mythic. Nietzsche had already suggested, in a not-so-covert attack on the modern theoretical man, that one could determine to what extent one was capable of understanding myth by asking himself if the miracles represented on the stage insult his historical sense, which insists on strict psychological causality, and whether he regards the miracle as a phenomenon intelligible to childhood, but alien to him.

> For in this way he will be able to determine to what extent he is capable of understanding *myth* as a concentrated image of the

world that, as a condensation of phenomena, cannot dispense with miracles.

Perhaps Rothko sorrowfully saw himself as the modern despaired of by Nietzsche, the "mythless man" who stands "eternally hungry, surrounded by all past ages and digs and grubs for roots, even if he has to dig for them among the remotest antiquities"—an image that could be, perhaps, corrected by art. Like Nietzsche, he turned to the earliest modes of Greek drama. It was Aeschylus to whom he returned; Aeschylus, in whom Nietzsche said the Apollinian and Dionysian were fused, and whose portrait of Prometheus might be expressed in the formula: "All that exists is just and unjust and equally justified in both." If we could imagine dissonance become man, says Nietzsche, "and what else is man?," this dissonance, "to be able to live, would need a splendid illusion that would cover dissonance with a veil of beauty." The early Greek dramatists understood this perfectly. Rothko felt it to be true. He had already indicated in the choice of his subjects his propensity for melancholy and his awareness of dissonance. Now he would move beyond the paralyzing consciousness of unjustness in the world. When Nietzsche tries to elucidate the pessimist's flaw, he reverts to Shakespeare, as did many other late 19th-century thinkers who became fascinated with Hamlet's ambivalence. Rothko followed Nietzsche. He maintained his interest in Shakespeare all his life. Nietzsche, in the famous passage that seemed to so many 20th-century readers an astonishing prophecy of Existentialism, says that the rapture of the Dionysian state leads to a chasm of oblivion that separates the world of everyday reality and Dionysian reality. As soon as "everyday reality" re-enters consciousness, it is experienced with nausea. In this sense, he says, Dionysian man resembles Hamlet for "both have once looked truly into the essence of things, they have *gained knowledge*, and nausea inhibits action; for their action could not change anything in the eternal nature of things; they feel it to be ridiculous or humiliating that they should be asked to set right a world that is out of joint."

In the face of such harsh truths, "man now sees everywhere only the horror or absurdity of existence . . . he is nauseated." Only art, Nietzsche proclaims, knows how to turn these nauseous thoughts about the "horror or absurdity of existence into notions with which one can live: these are the *sublime* as the artistic taming of the horrible, and the *comic* as the artistic discharge of the nausea of absurdity."

As Rothko himself was moving more and more toward a distancing from the "everyday" world, in which he had never felt thoroughly comfortable, much as he had longed to (his appreciation of Avery was explicit on that), he entered a psychological state that welcomed Nietzsche's bitter indictment of the "theoretical man" who seemed to dominate the modern world. In an almost unconscious movement, Rothko had, in his work of the mid-to-late 1930s, rejected even the modern tradition in visual art, a tradition in which he claimed to see a materialist bias, and a disregard for deeper sources. Later he would, through his works, explicitly endorse Nietzsche's insight, particularly in a curious passage where the young Nietzsche contrasts what he calls theoretical man with the artist. Like the artist, Nietzsche observes, the theoretical man finds an infinite delight in whatever exists. The artist, however, whenever the truth is uncovered, "will always cling with rapt gaze to what still remains covering even after such uncovering." The theoretical man, on the other hand, finds the highest object of his pleasure in an "ever happy uncovering that succeeds through his own efforts." Nietzsche's disdain for the happy, conscious drive of the analytic man, so unaware of the mysterious increment even in "everyday life," found answering emotions in Rothko. Rothko's later work could be seen as "what still remains covering." Nietzsche remained faithful to his youthful insight: in the late work, "Nietzsche Contra Wagner," he takes issue with those who "want by all means to unveil, uncover." He has learned "to stop courageously at the surface, the fold, the skin, to adore appearance, to believe in forms, tones, words, in the whole Olympus of appearance. Those greeks were superficial—*out of profundity* . . . Are we not, precisely in this respect, Greeks? Adorers of forms, of tones, of words? And therefore—*artists?*"[19] Nietzsche moved on from *The Birth of Tragedy* toward a more calm clarity, never forgetting, however, the residue of his glance into the abysm. So Rothko would also move. As he said "The progression of a painter's work, as it travels in time from point to point, will be toward clarity. . . ."[20]

The clarity Rothko desired was slow in coming. In the middle-to-

12. Untitled   c.1936

late 1930s he experimented restlessly with various voices, various approaches to his pictorial problems. His identification of music as a possible fundament for his expression was not easily translated on the canvas. There are hints about the importance of music in several sketches and gouaches of musicians, or people listening to music. Overtones of his meditation on tragedy also occur in the increasing flatness of his compositions. By reducing the depth of plazas, city-streets, interiors, Rothko simulates the narrow stage and enhances the illusion of backdrops, coulisses, and non-naturalistic props. Tragedy as a richly suggestive subject in itself, probably inspired by his reading of the ancient Greek dramas, begins to emerge sometime around 1936, in sketches and paintings of masked figures. One shows three frontal figures against a flat backdrop of a house façade. The upper figure suggests some reminiscence of Rouault's judges, but the two lower heads are the generalized masks associated with non-naturalistic drama. The central mask plays vaguely on the Janus theme (the left eye is drawn almost as a profile eye) that recurs in the "Antigone" possibly painted as early as 1938, and in a number of works of the 1940s. Was he thinking still of Nietzsche, who skillfully condensed his theory of the Apollinian-Dionysian doublet to the image of the Greek poet Archilochus juxtaposed with the head of Homer, side by side, on gems and sculptures? Homer, as Nietzsche wrote, "the aged, self-absorbed dreamer, the type of Apollinian naïve artist, now beholds with astonishment the passionate head of the warlike votary of the muses, Archilochus, who was hunted savagely through life." He adds that the modern interpretation would be that the first objective artist confronts the first subjective artist. The mask, the "covering," was to prove a central motif in Rothko's break with the figurative expressionist tradition.

The tempo of Rothko's explorations accelerated during the late 1930s. The autodidact was still at work, building his repertory of visual experiences, reflecting upon the nature of art, trying to reconcile his often contradictory impulses. He looked everywhere—not only to ancient sources of Near Eastern art, and Greco-Roman art in the halls of the Metropolitan Museum, but also at the works of the modern period beginning with the early 19th century. Joseph Solman specifically remembers his interest in Corot—a likely source

13. *Antigone*   1938-41

for Rothko to store in his memory. Corot, of whom Baudelaire said: *"toutes ses oeuvres ont le don particulier de l'unité, qui est un besoin de la mémoire,"* and who the modern art historian Germain Bazin echoes in: *"L'œuvre de Corot est donc une sorte de biographie de cette lumière."* It is possible that Rothko's increasingly lightened palette, and its frequent terra-cottas, yellows, and ochers during the last years of the 1930s was informed not only by Avery's corresponding lightening, but by his own appreciation of Corot's "biography" of the Roman light in 1827. There were also opportunities to see exhibitions of the work of Matisse, whose "The Red Studio" had been shown first in 1913 in the Armory show and had been widely reproduced before its acquisition by the Museum of Modern Art in the late 1940s. Matisse's condensations of time and space, which were to be of considerable moment to Rothko in later years, already attracted him—perhaps via Avery—while he was casting about for a likely idiom during the later 1930s. At the Metropolitan Museum he saw the Pompeian frescoes that would also come to possess his imagination in later years. They peer through some of the last figurative paintings. Then, there was Rembrandt, touching other chords in Rothko; Rembrandt with his eloquent manipulation of light, so much like staged drama; Rembrandt who may have inspired one of Rothko's earliest approaches to the drama of the Crucifixion (although Waldemar George had seen another source as more likely—the Italian Trecento).

As the 1930s approached the cataclysm that artists had been dreading for years, it was Miró who, for Rothko, epitomized the Nietzschean, rhapsodic vision of art. Miró was widely exhibited during those years; in the Museum of Modern Art's *Dada and Surrealism* show in 1936, and also at the Pierre Matisse Gallery during the same and following years. In 1941 there was a large retrospective at the Museum of Modern Art. Miró had accepted the revelations of Surrealism. Yet, his poetic temperament stood between him and the philosophic wing of the movement, and protected him from the verbosity and literary extravagances evident in so much of the Surrealist rhetoric. His attitude corresponded to Rothko's dawning realization that he could depart from the everyday world without sacrificing its vitality. There were many statements of Surrealist

14. JOAN MIRO   *The Hunter (Catalan Landscape)*   1923-24

theory published at the time, often in words of artists such as André Masson and Max Ernst, that could have lingered in Rothko's mind as he set to work to demolish his own past. But the statement of Miró, made in the work, was far more effective in spurring Rothko's turning. Miró's evocation of the mythic was itself informed by the excavations of Max Ernst, but Miró had been released from his earlier naturalistic illusionism when he discovered in himself the compelling emotional need for the poetic word. Over and over again he drew near to the poets, not only Eluard and Breton and Desnos, whom he met early in his sojourn in Paris, but also to Novalis, Rimbaud, Baudelaire, and, the poet among philosophers, Nietzsche. Miró specifically designated poetry as his inspiration and goal in the paintings he commenced in 1923. But he was a painter; a painter gathering his inner forces to make the leap that so often appears in his imagery—a leap into time, and into vastnesses such as sky and sea; into the hypnogogic realm posited by Poe and taken up by Baudelaire, in which images flow without causal interference, and time swells and shrinks with a mysterious pulse unknown to the wakeful mind. Breton's stress on automatism—thought's dictation without the interference of fixed mental structures—loosened up his writing colleagues, leaving them more accessible to their fantasies. But for painters, automatism was rarely adequate. Miró certainly dallied with automatism, which for visual artists amounted to exercise or doodling with pen or pencil in which the scribble technique was always linear. But he did not dally for long. (Rothko himself reported briefly experimenting with automatism in the late 1930s, recapitulating Miró's experience and, like the older master, moving on with his linear treasures to more complex experiences with paint.) Gathering up his references to specific images that had once peopled his paintings overtly, and mixing them with the symbolic fusions of imagery derived from the poetic diction of his favorites, Miró launched himself into a universe governed by his own imagination. The painting in which he most clearly eunuicated his new intention was "The Hunter," subtitled "Catalan Landscape," which entered the Museum of Modern Art's collection in 1936. If the translation of time into intercollated regions and simultaneities had attracted Rothko to the work of Eliot, it was projected into spatial

63

dimensions for him by Miró (who in turn was probably tutored, visually—at least a little—by what he saw in the studio of his neighbor, André Masson). The impact of this painting is apparent in Rothko's oeuvre for years after his first encounter with it at the museum. Miró had made a radical shift in which he found the material means to express his new vision. The means included the simple expedient of thinning his paint to almost transparent tones, suggesting the fluidity of his earth-sea-sky metaphor. The pale pink of the earth is so close in value to the yellows of sea and sky that the incursion of one on the other is easily assimilated by the eye. Scattered at various planes within the basic scheme are the symbolic objects, moored by no law, not even that of gravity. Rothko saw this painting long before meticulous art historical exegeses of the shapes and images were published. What he saw were ambiguous forms and less ambiguous forms treated gaily, and without emphasis, flying in every direction up and down the picture plane. The organization in three regions was cunningly subverted by the seeming disparateness of shapes. Mixed with allusions to organic creatures—forms such as eyes, spiders, human heads, and rabbits—were allusions to geometric figures: the geometer's triangle, a crested cone, spheres, and trylons. Explicitly there is a ladder, an image that so many modern artists had converted from its Christian connotations to the older and more suggestive tradition of Jacob. The painting, bathed in the irreal light—a light of late dawn or early dusk, the fabled golden hours—proposes a union of thoughts and images, a cosmos, that had an inherent magnetism for Rothko. Around the edges of his consciousness the cloud of the irreal was already touched by the golden light that unifies.

64

5

In those western European countries where democracy had been repudiated, wrote Stuart Davis, national chairman of the American Artists' Congress, in 1937, books were burned, paintings ridiculed and destroyed, music by great composers forbidden to be played, and the "right to studentship in the arts denied except within the barbed wire limits of a mystical and militaristic political censorship." Davis expressed the alarm and consternation of most artists who, having seen the ravages of the Depression in their own milieu, cast apprehensive glances at Europe and repeatedly asked themselves if something were not out of joint even in their aesthetic relationship to their time. Rothko, Gottlieb, Solman, and even mild-mannered Milton Avery were among the many who worried intensively during the late 1930s. They grappled with problems of art and society with an urgency induced by an all too vivid imagining of what it was like to be an artist in Italy, Germany, or Spain. Davis's strong position—that there was a new concept of the meaning of art and its role in society; that "art is one of the forms of social development and consciousness, which is in constant interaction with the other social forms in its environment"—was difficult to counter. Davis anticipated the riposte of his antagonists in a paragraph underlining the central problem:

> In contradiction to this conception of art as a particular but integral part of the social structure, there are some artists and

critics who prefer to think of art as an isolated activity having its own history which is determined by the talent of an aristocracy of genius throughout the centuries. Such a viewpoint limits the meaning of art to a matter of fine taste in certain relations of form and color and ignores the fact that these relations always refer to a concept of natural relations. These idealists observe that works of art have the most diverse subject matter: people, landscapes, still lifes, abstract spatial relations—and they conclude from this that subject matter has nothing to do with art. This is exactly the same as arguing that since we observe life in people, in trees, and in apples, that people, trees, and apples have nothing to do with life, and thus they look for life in the "form" of a disembodied essence, a "non-objective" life without material limitation.[21]

Davis, who always aligned himself with abstract artists, nonetheless drew back from the extreme limits of abstract expression (probably identified in his mind with the de Stijl group) and took the modified Marxist position that art could never transcend its historical circumstances. The notion of an "immanent" art history was derided even by so extraordinary a mind as that of Meyer Schapiro who, reading Alfred Barr's modest introduction to the 1936 exhibition, *Cubism and Abstract Art*, took him to task for "the pretension that art was above history through the creative energy of personality of the artist."[22] An expressionist artist who, like Rothko, had worked for years on the assumption that "spirit, expressiveness and personality" were the real meaning of art, was in an extraordinary quandary during the late 1930s, when increasingly his personal responses to nudes, trees, and apples seemed not only irrelevant but perhaps immoral. It was at this juncture that Nietzsche's exploration of the collective basis for universal expression provided an alternative. The nonhistorical bias of myth flew in the face of Marxist theory, but it also assuaged the social conscience of the perturbed artists who felt the weight of their time and wished to respond responsibly to the calamity.

Rothko continued to be exceptionally active in the civic affairs of artists and was clearly preoccupied with the moral conflict Stuart Davis had underscored. With his comrades he had picketed when the government seemed to be reneging on its commitments to the WPA; he had attended mass meetings and had experienced the solidarity his youthful ideals had envisioned. But he was no longer a raw youth. He was a man in his middle thirties, well aware of inherent conflicts even in the principle of union. For a time, a political radical who was also an artistic radical could make common cause with those taking tendentious positions. But there came a moment when the political and ideological struggles became too bitter. On April 4, 1940, the American Artists' Congress held a disastrous meeting during which it passed a resolution that to many passionately dissenting members appeared to sanction the Russian invasion of Finland. Immediately Meyer Schapiro, Ilya Bolotowsky, Lewis Mumford, Adolph Gottlieb, Stuart Davis, and Balcomb Greene resigned. Within days a statement appeared, signed by Avery, Bolotowsky, Gottlieb, Harris, and Rothko, among others, explaining the move and calling for a new organization. They accused the American Artists' Congress of implicitly defending Hitler's position by assigning the responsibility for the war to England and France, and failing to react to the Moscow meeting of Soviet and Nazi art officials and official artists "which inaugurated the new esthetic policy of cementing totalitarian relations through exchange exhibitions." The Congress, they said, no longer deserved the support of free artists. Two months later the Federation of Modern Painters and Sculptors was founded after many animated meetings to hammer out its goals. The formative meetings resulted in a concept of organization which would, or should, successfully avoid any artistic restrictions, but which could still maintain the old ethical positions of the 1930s that demanded of artists that they attend to social and political questions conscientiously. In the preamble to its constitution, the founders stated

> We recognize the dangers of growing reactionary movements in the United States and condemn every effort to curtail the freedom and the cultural and economic opportunities of artists in the name of race or nation, or in the interest of special groups in the community. We condemn artistic nationalism which negates the world traditions of art at the base of modern art movements.

From the beginning Rothko was committed to the Federation's program. His close friend Gottlieb was on the executive board, and Rothko was active on the cultural committee. He took responsibility for bringing in discussants, inspiring public forums, and looking for issues to which the organization should address itself. He took the educational role of the Federation seriously, urging widespread activities and publications, and exhibited for many years in their exhibitions. The diversity of the group was noted by the critics, who were generally pleased. Henry McBride wrote on May 22, 1942, in the New York *Sun* that "the requirements for joining the society apparently, are merely just not to be academic."

These outside activities seemed only to spur Rothko in his energetic drive to clarify the meaning of his own work; to "see the world as if for the first time." By 1938 he had begun to explore the pictorial possibilities of the Nietzschean vision. Hearing the two voices—the urgent voice of history and the distanced voice of myth—Rothko immersed himself in a struggle to talk about a subject with meaning, and at the same time, to talk as a painter talks. "Antigone" painted sometime before 1942 probably (possibly in 1938, as Rothko thought), is one of several paintings derived from the Greek bas-relief and other ancient sources. The masks in the top register are either chorus or gods, their terra-cotta color recalling Greek vases; the buttocks of the performers are in a middle ground, and the feet—both animalesque and human, true to the protean character of myth—rest on a stage-like pediment below. No question here but that the Greek tragedy was the source of his new stance. All this was accomplished in hesitant pale tonalities, often ochers and tans, with occasional Aegean blues or with terra-cotta grounds that insist on the Greek source. In these friezes of the late 1930s and early 1940s, Rothko initiates archaeological meditations in the form of horizontal divisions that formalist critics see as the source of his later spare abstractions with their two, three, or four levels or divisions. At the time they probably provided Rothko with the means to pictorialize his intuitions of layered time; of the metamorphic character of myth, and of the clear structure of Aeschylean drama (both on the formal and the psychological level; the chorus and the heroes and the gods and the people being clearly defined in a stylized dramatic format).

If Rothko was moving into another range of sensibility in the confines of his studio, he was not yet ready to expose himself to the scrutiny of the public. As a member of The Ten, he had exhibited extensively but had rarely won approval. Solman quotes Lou Harris who reported that Rothko smarted from the fact that his name was so rarely mentioned in the reviews, saying, "If only they would say Rothko stinks!" In 1940 he had an opportunity he accepted with alacrity. J. B. Neumann, whose opinion Rothko valued, offered to exhibit his work together with that of Solman and the French Expressionist Marcel Gromaire. It was at this time that Rothko decided to shorten his name from Marcus Rothkowitz to Mark Rothko. Rothko was still treading cautiously between his Expressionist preoccupation with moody figurative works and his new impulse to abstraction. Gromaire was much admired in Europe for his proletarian sympathies, described in grim brown-to-black visions of workers and peasants, powerful in their structure and superior to anything in the genre in the United States. In the exhibition in January 1940 Rothko apparently showed only the elegiac subway scenes and a few figure studies. His reticence continued that spring and the following in Federation shows, where he again exhibited the subway paintings. It wasn't until Sam Kootz, a progressive young dealer, organized a large exhibition at Macy's in January 1942 that Rothko publicly exhibited his Greek paintings "Antigone" and "Oedipus." That same year Peggy Guggenheim opened her Art of This Century Gallery.

During the Neumann show, Max Weber visited, and to Rothko's visible satisfaction warmly praised his work. Although Weber himself had long since moved from his early preoccupations, he remained faithful to the broad aesthetic precepts of his youth. Weber was still a firm believer in the mysterious power of the art of primitives. If Rothko began to pay close attention to Oceanic, African, Mesopotamian, and Archaic Greek visual arts, it was partly thanks to his early attention to Weber's teachings. As early as 1910 Weber was writing in Stieglitz's *Camera Work* about the superior expressiveness in the colors used by the Hopi in their katchina dolls, baskets, and quilts. Fifteen years later André Breton would call attention to native American art of the Southwest in his magazine *The Surrealist Revolution,* and fifteen years after that the Americans began to take cognizance of their own indigenous "primitive" arts as well as the primitive in other cultures.

In September 1940 Charles Henri Ford founded *View,* which was to be a lively extension of French Surrealist publications. The honored stars—Miró, Masson, Ernst—were discussed, illustrated, probed, and their "primitive" antecedents exposed. In 1942 there was an entire issue on Max Ernst. Surrealist interests were explored anew on American soil, sometimes by the old French band themselves who, escaping the war in Europe, foregathered noisily in New York. They brought with them their proudly brandished self-consciousness. The irony they injected—that the consciousness of self required an unconsciousness—was not lost on the New York artists who were still wrestling with their responsibilities in the wakeful world. The burden of consciousness was acknowledged by most of the New York painters who sought eagerly to unburden themselves. But they could not entirely banish their puritanical legacy. For most artists who would become New York School abstract painters, the Surrealist lunge into the unconscious was somehow shameless. What of the grand themes? Rothko, who may have remembered Clytemnestra's "Ah, it is sweet to escape all necessity'," was still troubled by the problem of meaning. He held a Bergsonian vision of the biological origin of consciousness, and understood the subterranean movements of the psyche to be somehow functions of a creative evolution. In one part of his soul, he was determined to express this psychological, yet historical, insight. In another, his encounter with the febrile modes of the painting Surrealists was for him a great release, a sweet escape. Surrealism was the elixir that could slake his thirst for revelation.

During the early 1940s, Rothko, Gottlieb, and their frequent companion, Barnett Newman responded actively to the presence of the Europeans. For several years they had despaired of their situation in a country that increasingly endorsed a nationalistic art pretending to epitomize a new aesthetic. Their disdain for American Scene painting, and realism in general, was total. Newman, sometime around 1942, still was fulminating against "isolationist art," pointing to a cunning strategy of the isolationist masterminds who, knowing the clock could not be turned back, agreed that modern aesthetic did not permit a re-emphasis of a subject matter. "And if subject matter isn't important, so what is? It's *how* the subject is painted. If it is a good picture it makes no difference what is painted."[23] Here he broached one of Rothko's most persistent themes: that manual skill was irrelevant; that "pictures" were not enough, and that the meaning of life was the real subject matter of an artist. There is some justice in seeing both Newman's and Rothko's defiance of French painterly cuisine in terms of their own shortcomings. Neither had ever had the rigorous training that could endow their fingertips with the unthinking skill to manipulate the elements of pictorial art. Newman had been particularly malhabile. Rothko had laboriously sought the means through countless experimental canvases. For him the indications in Surrealist art that a painter could use any means dictated by his subject matter were to be catalytic. The innate American suspicion of European cultivation seemed neutralized by the Surrealists' blithe renunciation of their own finesse, and their daring anarchy. Even so, the Whitmanian value of directness and innocent crudity always accompanied the American artists' bows in the direction of European culture. The attitude that wished to preserve American traits of independence had been a clear motivating force for American artists throughout their history. Whitman epitomized it when, in 1857, he wrote of Keats's poetry:

> Its feeling is the feeling of a gentlemanly person lately at college accepting what was commanded him there—who moves and would only move in elegant society reading classical books in libraries. Of life in the 19th century it has none, any more than statues have. It does not come home at all to the direct wants of the bodies and souls of the century.[24]

Rothko whose romantic anarchism had led him to bum in the loft manufactures of New York, and whose commitment to social justice had deep roots, was intellectually inclined to render justice to "the direct wants of the bodies and souls of the century." Temperamentally, he was better suited to the romantic vision of what Baudelaire called "the heroism of modern life." What he saw when Peggy Guggenheim opened her gallery was a succinct compendium of every modern idiom extant—not only Surrealist. Too thoughtful to evade the implications, Rothko charged himself with an arduous appraisal of the various stylistic responses to the modern world. If he was to navigate in a world that was patently out of joint, a world

15. *Horizontal Phantom*   1943

shattered by war and mired in the absurd, he wanted to be able to read his own compass. The "despair" of which so many painters spoke in 1940 (Newman: "In 1940, some of us woke up to find ourselves without hope"; Gottlieb: "During the 1940's, a few painters were painting with a feeling of absolute desperation"[25]) was not, as too many have assumed, only an aesthetic despair, but a despair born of events, *among which* one could include aesthetic events. Rothko sought to situate himself to be able to express the great clash of forces that could stand for what he knew. He begins with the Greek tragedies. His "Agamemnon" with its eagle rendered with a reminiscence of the American eagle of the Northwest Indians, combined with lips, claws, eyes drawn perhaps from Mesopotamian friezes, is a direct allusion to the fire of the classical world. Other paintings of the same period are filled with eyes, ears, and flames, as well as the iconographic symbols of medieval magic. Critics have seen both Masson's sinuous symbols of ecstatic man and beast, and Ernst's geologic references in works of the period, but these echoes are among many others. (For instance, America did not have to wait until Barnett Newman staged an exhibition of Pacific Northwest Indian art at the Betty Parsons Gallery in 1946. Kurt Seligmann, who settled in New York at the outbreak of the war, had already published an article in the 1938 edition of *Minotaure* on the totems and fetishes of the Haida.) Around 1942 Rothko painted a synoptic fusion of his imagination's sources. The feathered bird (eagle?) is decidedly American Indian. The eye is not Egyptian or Surrealist-dissected, but the eye of a biblical patriarch, or an Aeschylean elder, whose Expressionist hands are made to wring and whose cloud of beard is emphasized by the Ernst-like striations, made with the wooden handle of Rothko's brush. These Aeschylean overtones are sustained in many watercolors of the period. The fluid watercolor medium released the linear impulse with which the surrealists had limbered up their unconscious. In the oil paintings, which began to show distinct signs of Rothko's mounting assurance, not only in the greatly improved paint technique but in the handling of larger formats, the will to speak of the grand tragic themes is even more pronounced. In 1943 he painted "Horizontal Phantom," a fully orchestrated work with the newly lightened pal-

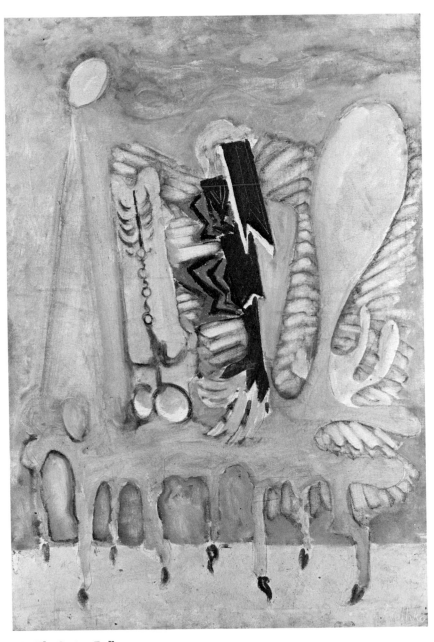

16. *The Syrian Bull*   1943

74

ette derived from Avery and Miró. Here are hints of Miró in the Merlin hat that cloaks a pink-legged figure on the right; hints of Picasso in the amphora-shaped figure prostrated (perhaps Iphigenia again being sacrificed); hints of Goya in the ghostly bird hovering in the sky; hints of American Indian totem poles in the phallic orange shape crested with an eye and bearing blood-color wings; hints of the Greek chorus in the gesture of hands that emerges from a water passage with waves rendered in a system of strokes very much as Avery characterized them. The whole is cast in three registers of time and space with movement suggested in the raking white lines scraped with the brush handle. The technical resourcefulness here can be attributed to Rothko's glances at other artists, among them the Surrealists from Europe and his own colleagues who, like him, were studying Klee, Kandinsky, and Picasso, among other 20th-century moderns. The subject matter—there had to be subject matter of significance for Rothko—was, as it was also in "The Syrian Bull" of the same year, meant to span human history, and to suggest its origins in the ancient Mediterranean basin—not only its origins in the human imagination, but its Darwinian origins, which would suggest the ground, the very matter from which consciousness, or rather imagination, would emerge. The welter of visual sources during the early 1940s, carefully documented recently in several scholarly analyses, were, in Rothko's case at least, adapted to his gathering awareness of a specific philosophic stance.

A public awareness of Rothko's awareness was about to become, for the first time, visible. In 1943 Rothko exhibited "The Syrian Bull" and Gottlieb "The Rape of Persephone" in the Federation's Third Annual exhibition, presented with a statement by the organizers that "we are now being forced to outgrow our narrow political isolationism" in order to recognize "cultural values on a truly global plane." Since most of the Federation's members were interested in developing a respectful public for their art, they actively tried to engage press critics in a public dialogue. Edward Alden Jewell, who was a shrewd journalist and knew the value of controversy, gladly fell in with the educational aspirations of the Federation. In a long review of the 1943 Annual, he wrote of his "befuddlement" over Gottlieb's painting, and his bafflement over Rothko's

"Syrian Bull." A few days later he reviewed his review and announced to readers that the Federation would reply to this befuddlement. And, a few days after that, June 6, 1943, he returned again to the show and the "enigmas" of Rothko and Gottlieb, this time reproducing both paintings. He then on June 13 reproduced sections from a June 7 letter from Rothko and Gottlieb. He must have gratified the cultural committee of the Federation when he suggested that the Federation might be spawning a new movement called "globalism."

There are various accounts of how the famous letter came to be written, but judging from an extensive series of drafts in Rothko's papers at the Rothko Foundation, it would seem that he first had the idea of answering Jewell's provocative remarks on his own behalf. After many drafts he probably checked with Gottlieb who volunteered to collaborate and promptly telephoned Jewell to offer the joint letter for publication—an opportunity Jewell wisely accepted. The original drafts with their minute variations indicate how strongly Rothko felt about defining his mission. He struggled to find the precise locutions, and laid stress upon the background for his position in modern thought. Thoughts echoed in the final letter appear in Rothko's notes, such as:

> Why the most gifted painters of our time should be preoccupied with the forms of the archaic and the myths from which they have stemmed, why negro sculpture and archaic Greeks should have been catalyzers of our present day art, we can leave to historians and psychologists. But the fact remains that our age [is] distinguished by its distortions, and everywhere the gifted men whether they seat the model in their studio or seek the forms, all have distorted the present to conform with the forms of Ninevah, the Nile or the Mesopotamian plain. . . . To say that the modern artist has been fascinated by the formal aspects of archaic art is not tenable. Any serious artist will agree that a form is significant insofar as it is expressive of that noble and austere formality which these archaic things possess. . . .

> I am neither the first nor the last compelled irretrievably to deal with the chimeras that seem the most profound message of our time. . . .

In another draft, Rothko again links the return to ancient motifs to the modern sensibility:

The truth is therefore that the modern artist has a spiritual kinship with the emotions which these archaic forms "imprison" and the myths which they represent. The public therefore which reacted so violently to the primitive brutality of this art reacted accurately, more truly than the critic who spoke about forms and techniques. . . .

And in still another draft, Rothko alludes to the condition of the period, and his shock in the face of the terrible public events:

Perhaps the artist was a prophet and many decades ago discovered the unpredictability which lay under man's seemingly ascending reason, or saw the potentiality for carnage we know too well today . . . the artist paints what he must . . .

There are still artists who still confound sunset and long shadows; the melancholy aspects of the times of day with the tragic concepts with which art must deal; but in all the art which jolts, moves and instigates to new discoveries is the art which distorts . . .

These drafts, with many erasures, recapitulations, penciled hesitations, and exclamations, speak of Rothko's continuing meditation on the grand tragic themes. His reading in the Russians, particularly Dostoyevsky and Tolstoy (whose *The Death of Ivan Ilyich* was one of Rothko's strongest incitements—he mentioned often the impact this stern and uncompromising *memento mori* had had on him), is reflected in the constant allusion to the irrational in man and his tragic position. When Rothko joined forces with Gottlieb, the letter became more assertive, marked less by Rothko's melancholy turn of mind.

Seizing the opportunity to formulate their principles in a public forum, Rothko and Gottlieb, with editorial advice from Barnett Newman, who may have written the first paragraph of the letter, warmed to the task. They allowed themselves boyish extravagance and heavy-handed sarcasm. The anarchist instinct to belabor authority came out strong in Rothko (whose ideas may have dominated here, as certain friends have insisted). He, in turn, was probably cheered on by the redoubtable anarchist Newman. As a point of honor the authors begin with a sarcastic attack on the workings of the critical mind, "one of life's mysteries." They chortle in rather sophomoric tones that it is "an event when the worm turns and the critic of the times quietly yet publicly confesses his 'befuddlement,' that he is 'non-plussed' before our pictures at the Federation Show." They then declare that they do not intend to defend their pictures, which they consider clear statements—the proof of which is Jewell's "failure to dismiss or disparage them." This the painters regard as "prima facie evidence that they carry some communicative power." They do not intend to defend their paintings, but *not* because they cannot—an "easy matter" to explain, as Rothko points out concerning "The Syrian Bull," which he calls "a new interpretation of an archaic image, involving unprecedented distortions." "Since art is timeless, the significant rendition of a symbol, no matter how archaic, has as full validity today as the archaic symbol had then." He adds, with an arrogance no doubt donned for the occasion, "Or is the one 3,000 years old truer?"

In the fifth paragraph, the painters settle down to serious business. No possible set of notes can explain our paintings, they declare, because

> explanation must come out of a consummated experience between picture and onlooker. The appreciation of art is a true marriage of minds. And in art, as in marriage, lack of consummation is ground for annulment.

This was, and remained, one of Rothko's most cherished principles. (Rothko is definitely the sole author of the fifth paragraph.) Its long life in the history of aesthetics was known to him, and known specifically through his careful reading of Nietzsche who in turn had carefully read Schopenhauer. Theories of the potential communion between artist and receiver had floated down the mainstream of modern thought from the Romantic period onward. Of all the theories of aesthetics, the one that appealed most to Rothko's imagination was one that insisted on the reciprocity of artist and viewer, of artist and the world—a theory of empathy that had been favored since the 19th century. Implicit in 19th-century attitudes shaped by the Romantic spirit was a belief in the dynamic process initiated by nature within the artist and transmitted to others who, in their own imaginations, complete the process.

78

After their arch preliminaries, Rothko and Gottlieb spell out their program soberly, producing one of the few manifestoes of the Abstract Expressionist era. Rothko's allegiance to Nietzsche is apparent in the first point:

> 1. To us art is an adventure into an unknown world, which can be explored only by those willing to take risks.

His yearning for release from necessity appears in the second and third points:

> 2. This world of the imagination is fancy-free and violently opposed to common sense.
> 3. It is our function as artists to make the spectator see the world our way—not his way.

In the fourth point the artists revert to a belated recognition of long-held assumptions of modern artists:

> 4. We favor the simple expression of the complex thought. We are for the large shape because it has the impact of the unequivocal. We wish to reassert the picture plane. We are for flat forms because they destroy illusion and reveal truth.

In the fifth and most important point, Rothko declares once again, as he had already hinted in the *Whitney Dissenters'* catalogue, and in his notes for the Brooklyn Academy speech, that mere manual skill is not enough; that the high seriousness of the artist goes beyond technique and decoration. Again, Nietzsche's characterization of Dionysian man who feels it ridiculous or humiliating that he be asked to set right a world that is out of joint underlies the statement. Nietzsche's solution—that only the artist can turn the horror or absurdity of existence into notions with which one can live—seems to be the one Rothko is proposing with the statement:

> 5. It is a widely accepted notion among painters that it does not matter what one paints as long as it is well painted. This is the essence of academicism. There is no such thing as good painting about nothing. We assert that the subject is crucial and only that subject matter is valid which is tragic and timeless. That is why we profess spiritual kinship with primitive and archaic art.

This general statement of principle was amplified five months later on October 13, 1943, when Rothko and Gottlieb broadcast on WNYC at the invitation of Hugh Stix, whose activities included running a gallery for experimental artists. It is clear that the text for the broadcast was carefully worked out in advance, and that the two painters set great store by the lucid enunciation of their principles. The tone reflects, certainly, their excitement, both at the attention they had received since the *New York Times* coverage, and their own rapid evolution in their work. The immediate occasion was an exhibition of portraits by Federation members, "As We See Them," in which Rothko showed a painting called "Leda" and Gottlieb showed his "Oedipus." Gottlieb introduces the subject by posing four questions he claimed were asked by a correspondent, asking Rothko to answer the first: Why do you consider these pictures to be portraits? Rothko responds that there is

> a profound reason for the persistence of the word "portrait" be-
> cause the real essence of great portraiture of all time is the
> artist's eternal interest in the human figure, character and emo-
> tions—in short in the human drama. That Rembrandt expressed
> it by posing a sitter is irrelevant. We do not know the sitter but
> we are intensely aware of the drama. The Archaic Greeks, on
> the other hand used as their models the inner visions which they
> had of their gods. And in our day, our visions are the fulfillment
> of our own needs. . . . What is indicated here is that the artist's
> real model is an ideal which embraces all of human drama rather
> than the appearance of a particular individual.

> Today the artist is no longer constrained by the limitation that
> all of man's experience is expressed by his outward appearance.
> Freed from the need of describing a particular person, the pos-
> sibilities are endless. The whole of man's experience becomes
> his model, and in that sense it can be said that all of art is a
> portrait of an idea.

Gottlieb then addresses himself to the next question: Why do you as modern artists use mythological characters? He asserts that artistically literate people have no difficulty grasping the meaning of Chinese, Egyptian, African, Eskimo, Early Christian, Archaic Greek, or even prehistoric art, and that he and Rothko use images that are directly communicable to all who accept art as the language of the spirit. The third question—Are not these pictures really abstract paintings with literary titles?—gives Rothko the opportunity to frame more explicitly his rejection of the formal modern tradition. Their paintings, he says, are not abstract paintings since it is not their intention either to create or to emphasize a formal color-space arrangement:

> If our titles recall the known myths of antiquity, we have used them again because they are the eternal symbols upon which we must fall back to express basic psychological ideas. They are the symbols of man's primitive fears and motivations, no matter in which land or what time, changing only in detail but never in substance . . .

> Our presentation of these myths however, must be in our own terms which are at once more primitive and more modern than the myths themselves—more primitive because we seek the primeval and atavistic roots of the ideas rather than their graceful classical version; more modern than the myths themselves because we must redescribe their implications through our own experience. Those who think that the world of today is more gentle and graceful than the primeval and predatory passions from which these myths spring, are either not aware of reality or do not wish to see it in art. The myth holds us, therefore, not thru its romantic flavor, not thru the remembrance of the beauty of some by gone age, not thru the possibilities of fantasy, but because it expresses to us something real and existing in ourselves, as it was to those who first stumbled upon the symbols to give them life.

Gottlieb then answers the last question: Are you not denying modern art when you put so much emphasis on subject matter? He states that it is generally felt that the emphasis on the mechanics of picture-making has been carried far enough, and that though the Surrealists had asserted their belief in subject matter, for him and Rothko, "It is not enough to illustrate dreams." Gottlieb continued, stressing the particular situation of artists in 1943:

> If we profess a kinship to the art of primitive men, it is because the feelings they expressed have a particular pertinence today. In times of violence, personal predilections for niceties of color and form seem irrelevant. All primitive expression reveals the

constant awareness of powerful forces, the immediate presence of terror and fear, a recognition and acceptance of the brutality of the natural world as well as the eternal insecurity of life.

These views, as romantic as they were, were shared to some degree by other artists of the New York School. A tone of high seriousness had crept into artistic discourse, probably induced as much by a new-found independence as by awareness of the savagery at loose in the world. Rothko and Gottlieb by 1943 were circulating among artists who were feeling excitement in the emancipation of their energies and enjoying a new climate of encouragement. Not the least important in the rapid unfolding of a new vision among American painters was the existence of Peggy Guggenheim's gallery, in which the American aspirants could rub shoulders with stellar Europeans. From 1943 to 1944, the Art of This Century Gallery offered a spate of vanguard American exhibitions, starting with Jackson Pollock's first one-man show in November 1943, introduced by James Johnson Sweeney. He called Pollock's talent lavish, explosive, and untidy, and said that "what we need is more young men who paint from inner impulsion." It was reviewed by Clement Greenberg in *The Nation* enthusiastically, and also by his younger colleague Robert Motherwell in *Partisan Review*, who called Pollock "one of the younger generation's chances." The 1944 season began in October with William Baziotes's first one-man exhibition, followed in late October by Motherwell, and in January 1945, by Rothko.

Rothko's consideration of portraits as ideals that embrace all of human drama was not mere lip service. From around 1943 he had plunged into an atmosphere of perpetual drama and was undergoing the emotional excitement sometimes described in relation to religious conversion. (William James, describing the conversion experience in *The Varieties of Religious Experience*, wrote: "All we know is that there are dead feelings, dead ideas, and cold beliefs, and there are hot and live ones; and when one grows hot and alive within us, everything has to re-crystallize about it.") Rothko's excitement, his unabated attention, his new-found spiritual release, his exhilaration on having, at last, taken the risk Nietzsche had advocated can be felt in the flow of work. He was expanding all his horizons. In the summer of 1943 he had traveled to California, where he first encountered the already notable rebel painter Clyfford Still briefly. In the late fall of 1944 after his separation from his first wife, he met Mary Alice Beistle, whom he married the following March. While he was courting her he gave her a copy of Kafka's *The Trial*. It had appeared in English in 1937 and had ever since been discussed in the little magazines with mounting interest (so much so that by 1947 Edmund Wilson complained that the American literati had produced a cult). Rothko's interest in Kafka was natural enough. He had been a familiar of Kafka's sources in Russian literature, particularly Dostoyevsky; and also of Kierkegaard. As a Jew transplanted into a non-sectarian culture, Rothko was sensitive to Kafka's paradoxes. The entrance to the place of judgment that Joseph K. so fruitlessly sought was, as Rothko said about his own doorways, interpreted as the point at which the artist left the world in which plans and ideas occurred. The parallel world Kafka described, with its deep perspectives giving way to shallow stage sets, and its strange inversions of time and space, was precisely what Rothko was bent on expressing in the works he showed at the Art of This Century. These works represented how he wished to be understood by the world (and how he wished to understand the world, for his views of the nature of art were based on what philosophers had begun to call "intersubjectivity" in the 20th century). They were how he wished to be understood by his fiancée. If he was struck by Kafka's *The Trial*, he was struck by the way Kafka clung to the material world and yet, through distortion, even visual distortion in his strange description of mirrored palaces, endless corridors, simultaneously perceived places of different orders, presented a texture of a world immediately identifiable as mythic; a world in another register, a kind of basso continuo to what Nietzsche called "everyday" life.

The paintings in the first really significant one-man exhibition in Rothko's life, at the age of forty-two, reflected his preoccupations, not only in their suggestive forms, but in their titles as well. On one side there was the Aeschylus atmosphere conjured in the 1942 "Sacrifice of Iphigenia" with its totemic shape, its supplicating hands and robed deity—the transformed Merlin hat. Also "Tiresias," with its emphasis on the seer's eye, and "Hierarchical Birds," its pale

17. *Hierarchical Birds*   c.1944

blues over pinks calling up an Aegean drama; its three divisions offering a frieze effect, and its brushed lines—scallops and chevrons forming a sign language very much like Avery's—evoking the Aeschylean eagle's lament. And again, "The Syrian Bull," "Omens of Gods and Birds," and "Ritual." On the other side there was a group of works reflecting Rothko's preoccupation with a Darwinian vision of the emergence of consciousness, among them "The Birth of the Cephalopods," its thinly layered pinks and golds symbolizing the primordial dawn; and "Slow Swirl by the Edge of the Sea," listed in the catalogue as "Stars and Swirl By at the Edge of the Sea." Recent critics have alluded to Rothko's studies in geology at Yale to account for the layered vision and fossilized forms in "Slow Swirl." They have also suggested that his close relations with Barnett Newman influenced his choice of subject matter. It is possible that in passing, Rothko drew upon his rudimentary knowledge and brief exposure to the natural sciences, and that Newman's keen interest in science, especially ornithology, might have fused with numerous other inflowing interests at this time. It was a time in Rothko's life when the creative afflatus was powerful. All kinds of images and impulses found their way into the paintings that multiplied at an unprecedented rate. In the works exhibited, there were various climates and two main approaches. In "Birth of the Cephalopods" and "Slow Swirl by the Edge of the Sea" Rothko translates his intuition of beginnings through fluidity and the implicit metamorphic character of his arabesques. Here, all is "becoming"—far from scientific causes and effects. In these newly fluent visions, Rothko has almost disembarrassed himself of necessity. The canvas is the one place where such freedom, such an extravagant dream is possible for him. It is the beginning of his exodus from the everyday world, despite his protest that the earthworm was his source. The cephalopods transpire in a spaceless environment peculiar to the dream. In "Slow Swirl"—a painting engraved in the memories of many viewers and justifiably considered a capital summary of the period—the color is etiolated, diffuse: an analogue to an imagined realm where qualities alone, and not the weight of matter, exist. It is the no-color of the sea. The color with no name that so many painters had strived to record, to re-live on canvas, which the sheer matter of oil paint itself has so often defeated. It is as abstract as the *clarté* of which Baudelaire spoke in so many unaccustomed contexts. This triumphal exit from "the world of ideas and facts"; this approach to the "doorway" is the single most significant aspect of the painting. Rothko created his very own light, in the light of which all experience would henceforth be filtered. Within this ambiance there might well be allusions to biologic nuclei, paramecia, surrealist bones, but far more important is the discernible reference to Rothko's secret impulse to rival the art of the musician. There are arabesques, spirals, and clusters of bars that can be read as musical symbols. In the main shapes, the two gyrating figures (that admittedly bear comparison to Ernst, Dalí, Masson, and his own colleague Jackson Pollock) have vestigial feet—an overt reference to the dance. The existence of clef signs and the scrollwork of violins unmistakingly refers to the reception of music, and the clusters of linear flourishes produce movements with distinct rhythmic intentions as their lines intersect. The axis of each figure is vertical, and the figure itself can be understood as a play on the arabesque, an infinity sign.

During the same year Rothko began to work out a slightly different exodus route from the given world of things and events. "Poised Elements" and one or two other works such as "Olympian Play" are characteristic. Here, Rothko defines more precisely the light that could never be associated with the hard objects picked out by the "everyday" light of day. He has discovered the light within the canvas itself, as Miró had before him. He scrapes the surfaces in "Poised Elements" to uncover their inmost threads, and he divides his composition into three zones of experience, inter-related by the rather sharply defined forms. These tiers of reds, grays, greenish-grays shelter certain specific references—to a chrysalis perhaps, and to birds and shadows. Above all, they suggest the mirrored world of interior fantasy; a light of otherness that is the beginning of Rothko's conquest of antinomies. From now on there will be many allusions to white and whiteness as a kind of nimbus, an aura, a natural enough allusion to the ancients who imagined a luminous vapor or cloud hovering about their gods who visited earth.

In this group of works of roughly 1944-46, Rothko is emboldened to use accents of dramatic color, most often red, that bear associa-

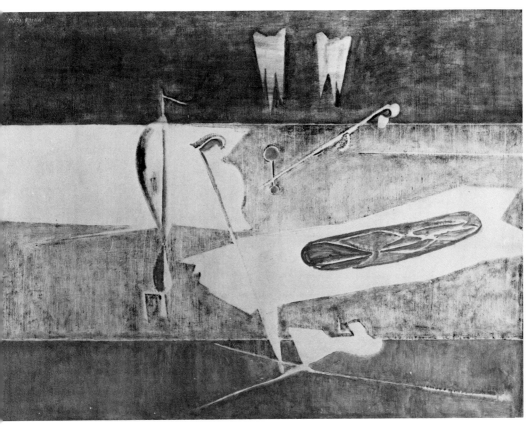

18. *Poised Elements* 1944

19. *Olympian Play*   c.1944

tions with primitive ritual. He also for the first time experimented with saturated blacks, using them to symbolize recession. In this he may have been encouraged by the works of Jackson Pollock exhibited in 1943 at the Art of This Century. Pollock had thrown out the old Impressionist caveat, and was dramatically underscoring his own mythic paintings with liberal areas of harsh blacks. For Pollock, black certainly represented emotional power; for Rothko, darkness and fertility, earth and night.

These paintings in Rothko's first one-man show at Art of This Century in 1945 were introduced with an unsigned foreword (by Peggy Guggenheim's assistant Howard Putzel?) that almost certainly reflected Rothko's own evaluation of his work at the time. Rothko was in the habit of trying to control everything connected with his work, even the written commentary. Elaine de Kooning describes a bizarre encounter a few years later in which Rothko, given an opportunity to read over an article before publication, completely rejected it. When de Kooning offered to re-write it, he insisted on accompanying her to her studio immediately, where he remained all night as de Kooning revised her text. Rothko passed on it page by page between beer and pastrami sandwiches. For a catalogue foreword to what he regarded as his most important exhibition, Rothko would have been no less zealous. "Rothko's style has a latent archaic quality," the writer of the Art of This Century preface asserts. "This particular archaization, the reverse of the primitive, suggests the long savouring of human and traditional experience as incorporated in the myth." We can hear Rothko in the following:

> Rothko's symbols, fragments of myth, are held together by a free, almost automatic calligraphy that gives a peculiar unity to his paintings—a unity in which the individual symbol acquires its meaning, not in isolation, but rather in its melodic adjustment to the other elements in the picture. It is this feeling of internal fusion, of the historical conscious and subconscious capable of expanding far beyond the limits of the picture space that gives Rothko's work its force and essential character. But this is not to say that the images created by Rothko are the thin evocations of the speculative intellect; they are rather the concrete, the tactual expression of the intuitions of an artist to whom the subconscious represents not the farther, but the nearer shore of art.

This statement contains the contradiction that Rothko and others at the time most worried over: the primacy of intuition versus the importance of art as knowledge. Rothko's position was, and would remain, that such antinomies must not be resolved, but must be held in suspension. The writer's last sentence, abjuring the "speculative intellect," was a reflexive gesture of self-defense that most American artists made in the face of the critics who were hostile on both counts.

Rothko's show, that he had hoped would find, finally, critical acclaim, did not make much of a stir. The daily newspapers, that had so recently afforded him so much gratification, ignored it. In the art journals he was favorably but briefly reviewed. The reviewer in *Art Digest* (Jan. 15, 1945) remarked that "Mark Rothko has been a kind of myth in contemporary art for about ten years" and spoke of "the extent of courage a painter must have to make a sortie of this kind into unexplained territory." Rothko, knowing that the state of reviewing at the time required that he explain his unexplained territory, composed a personal statement for an exhibition a month later at David Porter's Gallery in Washington, in which Pollock, Baziotes, and Gottlieb were among the exhibitors. Rothko opened with his familiar insistence that he adhered to the material reality of the world and the substance of things. "I merely enlarge the extent of this reality, extending it to coequal attributes with experiences in our more familiar environment":

> I insist upon the equal existence of the world engendered in the mind and the world engendered by God outside of it. If I have faltered in the use of familiar objects, it is because I refuse to mutilate their appearance for the sake of an action which they are too old to serve; or for which, perhaps, they had never been intended.

Rothko goes on to make a careful distinction that shows he had recourse to philosophical currents then being discussed in New York, currents that would soon become obvious in the Existentialist vocabulary adopted even by the art critics:

> I quarrel with surrealist and abstract art only as one quarrels with his father and mother; recognizing the inevitability and

function of my roots, but insistent upon my dissension; I, being both they, and an integral completely independent of them.

Rothko separates himself from Surrealism, as he had in 1943 statements, on the basis of humanism—the humanism that acknowledged, or rather, fed upon the culture of the ancients in the tragedies. The Surrealists, he said, had established a congruity between the phantasmagoria of the unconscious and the objects of everyday life. "This congruity constitutes the exhilarated tragic experience which for me is the only source book for art." But, he added:

> I love both the object and the dream far too much to have them effervesced into the insubstantiality of memory and hallucination. The abstract artist has given material existence to many unseen worlds and tempi. But I repudiate his denial of the anecdote just as I repudiate the denial of the material existence of the whole of reality. For art to me is an anecdote of the spirit, and the only means of making concrete the purpose of its varied quickness and stillness.

He ends on a resounding note:

> Rather be prodigal than niggardly I would sooner confer anthropomorphic attributes upon a stone, than dehumanize the slightest possibility of consciousness.[26]

This statement was not to be superseded by later statements or even by the seeming abandonment of subject and symbol in his own abstract works. Even after he left behind apparent anecdote, it remained harbored in his imagination. A few years later he was telling students to avoid French-style abstraction and abstracting from the model: "I'd rather paint eyes on a rock, put a couple of eyes on a rock," he told them, "than take the human figure and make it mechanical the way Picasso does. Or Léger."[27]

6

During the 1940s Rothko's life was punctuated with events that sustained his excitement and incited his daring. His intensifying awareness of himself as an individual "I, being both they [roots], and an integral completely independent of them" became evident to friends who noted Rothko's increasing recourse to polemic. Both privately and publicly he harrowed away in the subsoil of his psyche. His need to formulate a philosophy of art was shared by many of his painting colleagues who, after the Second World War, burst into a new round of excited discussion that was to lead to a vague grouping later dubbed Abstract Expressionism. Rothko, always known for his probing conversation and his ability to wield various theories in the Socratic manner (many of his old friends agreed that he could have been a lawyer or a clergyman), redoubled his effort to clarify his own stance. Having already found the Surrealist position wanting, he was open to many alternatives. One of them was presented by the North Dakota-born painter Clyfford Still, whom he had first met in San Francisco in 1943. Still, who was forging a colorful myth about himself even in those days, exercised a powerful attraction for Rothko, the pensive descendant of bookish people. Still was the very personification of the New World pioneer: self-reliant, defiant, pitting himself against nature. He shared the new continent's suspicion of European culture, and was, at least in the myth he proposed for himself, firmly opposed to any intellectualization of art. He talked all the time about "energy and intuition" and about the need to get out from under decadent European influence. He was also given to boasting about his virile youth, in Alberta, Canada, where "my arms have been bloody to the elbows shucking wheat" and where he would take a horse at night for five miles "to bang out Brahms and look at art magazines."[28] Still's fierce independence and his hatred of anything that smacked of "collectivism" impressed Rothko. Still became Archilochus to Rothko's Homer. With his brash rhetoric (interlarded with terms drawn from baseball rather than aesthetics) he both amused and fascinated Rothko. Even early on Still was given to vivid tirades at which he would excel in later years. Motherwell has called him the John Brown of the art world. Rothko, who had never forsaken his anarchist rebelliousness, took Still seriously. When Still arrived in New York in the late summer of 1945, Rothko visited him in his studio in Greenwich Village without being invited (according to Still in his later account thoroughly tinctured with malice). There Rothko saw some of the easel-sized paintings done the year before in Virginia, many of them dark, roughly painted images in which wrench-like or totemic shapes were dramatically lighted in a kind of Götterdämmerung atmosphere. Although Still's sensibility was entirely different from Rothko's, Rothko responded to the drama these canvases undeniably embodied. He reported back to Peggy Guggenheim enthusiastically and persuaded her to visit Still's studio shortly after. She decided to exhibit Still in her autumn opening group show. Soon after, she offered him a one-man show for the following February. In the interim Rothko and Still drew together, although Rothko's other close associates, Gottlieb and Newman, had some reservations about this arrogant newcomer. Rothko seemed to admire Still's arrogance, and Still, despite his later disparagement, decidedly encouraged Rothko's devotion. When the time came for the Art of This Century show, it was Rothko who wrote the catalogue, and we can hardly credit Still's later version that he had not wished Rothko to do it, and that Rothko had used him for his own purposes. Still probably encouraged Rothko's interpretation that tended to identify him with the Art of This Century group (Baziotes, Pollock, Motherwell, Rothko especially), whom Rothko called the "small band of Myth

20. Untitled   1945

94

Makers who have emerged here during the war." Rothko saw Still expressing the "tragic-religious drama which is generic to all Myths at all times, no matter where they occur" and, with generous intention, declared that Still was creating "new counterparts to replace the old mythological hybrids who have lost their pertinence in the intervening centuries." Finally, Rothko alluded to the Persephone myth and quotes Still himself as saying that his paintings are "of the Earth, the Damned and of the Recreated." Years later, what Rothko remembered writing about Still was about the primal matter from which these paintings were wrung—the environment of the earthworm.

Rothko's encounter with Still's paintings was as fruitful as his encounters with works of other contemporaries from whom he learned. He, who had struggled so mightily to master his means, was heartened to observe, as he had already to some degree discovered in the works of the Surrealists, that a nice technique, such as the French painters seemed to command, was not at all necessary to expression. Still couldn't care less about conventional techniques. He trowelled on his paint with indifference to modulations, concerned only with keeping alive his intuition and displaying his energy. His melodramatic propensity for sharp light effects was evidence that, like Rothko, he thought of painting as a counterpart to theater.

When, in April, Rothko had an exhibition of his watercolors and gouaches arranged by Betty Parsons at the Mortimer Brandt Gallery, the incursion of Still's earthy, damned, and re-created vision in Rothko's imagination was beginning to be visible, but it wasn't until his exhibition of oils at the San Francisco Museum of Art the following August that a distinct note of Still's turbid surfaces is felt in a few paintings, such as "Untitled," in which Rothko uses far more restless shapes than is usually his wont, and where Still's monkey-wrench stylizations as well as his zig-zagging eruptions of light in darkness appear. However, there are familial likenesses in these paintings to Pollock and Motherwell just as much as to Still.

Even Still was susceptible to the gathering group consciousness in the painters' quarters in those days. Painters met on neutral ground, such as the Waldorf Cafeteria, or, in the later 1940s in a few favored taverns, such as Minetta Tavern on MacDougal Street, and later the Cedar Tavern on University Place, and thrashed out their views of what they perceived to be a new situation. Rothko, who did not live downtown and was considered by some artists to be something of a loner, all the same made frequent sorties to find his confreres in the downtown bars. He maintained friendly relations with a number of artists and visited studios regularly. One of the artists who caught his attention during that highlighted season when he had his first important one-man show was the young Motherwell, fresh from serious studies in literature and art history, and eager to test his burgeoning theories with older artists. Rothko liked to talk to Motherwell, and, as in the case of Still, Motherwell probably represented a refreshing change. "I was Stephen Dedalus to his Bloom," Motherwell says. "We were exotic to each other." They had met during the afternoon when Rothko was hanging his show. That spring they saw each other from time to time, and during the summer of 1946, when Rothko had rented a small cottage in East Hampton, they saw each other frequently, dining together at least once a week and having extended conversations. When they returned to New York, Rothko introduced his young friend to his circle, among them Barnett Newman, Herbert Ferber, Adolph Gottlieb, and Bradley Walker Tomlin.

In 1944 Motherwell had given a talk at Mount Holyoke College, "The Modern Painter's World," in which he had addressed himself to questions that were regarded by New York painters as extremely significant. This was published in *Dyn*, a review published in Mexico City and widely read by New York artists. In his article, shaped within the framework of Marxist analysis that he had learned in his studies with Meyer Schapiro during the late 1930s, Motherwell takes cognizance of historical oppositions between painters and the bourgeoisie, but suggests that the experience of the artists who had known Spain, Hitlerian genocide, and the carnage of the Second World War had mitigated the old arguments. The problem for the emergent artist *"is with what to identify himself."* Since in Motherwell's view the middle class was decaying and the working class had forfeited its consciousness, there seemed little alternative for painters but to paint for each other. He too, like Rothko, Newman, Still, Baziotes, and Gottlieb among others, had outgrown the dogmas of Surrealism and specifically rejected their disdain for "the mind" or formal values in painting. He had concluded that painting is "mind realizing itself in color and space. The greatest adventures, especially in the brutal and *policed* period, take place in the mind."

Rothko's mind was working furiously in the latter half of the 1940s. The flow of his work was unabated, and his reflection on its sources gained in depth as he, like so many others at the time, engaged in spirited discussion. The evidence abounds in periodicals and letters of the late 1940s that artists had crept out of their proverbial isolation in American society and were establishing themselves in the intellectual community as valid cultural figures, much to the bemused surprise of a number of literary figures who could never quite believe that this caste of handworkers had much to contribute. There were, however, several members in the artists' milieu who were determined to raise the status of the visual artists via the literary magazines. Harold Rosenberg, who had a foot in both camps, was busy with Motherwell, plotting a new, intellectual artists' publication, while Ruth and John Stephan were readying their quarterly, *The Tiger's Eye*. Both journals appeared in the fall of 1947. *Possibilities*, the Rosenberg-Motherwell project, was destined to have only one issue, but it was an issue brimming with questions and inquiries, brilliantly summing up the preoccupations of the immediate postwar art world. Its most salient attitude was one of disaffection with ideology and politics. These prewar concerns of artists were considered by almost everyone to have been fruitless and naïve. Writers, as Isaac Rosenfeld had noted as early as 1944, who were searching for a "belief broader and more reliable than politics regard this as their only alternative: to go forward by going back into themselves. And even if this is not the only means for making themselves 'men of substance' still they will have found a different means. . . ."[29] Rosenberg, moving in the same circles as Rosenfeld, would be the agent through which this attitude became validated for visual artists. Together with Motherwell, he gave a precise profile to a growing conviction that swept past ideologies and aesthetics, into the unknown. The confluence of streams of thought during the last few years of the 1940s is remarkable. Among the forceful voices emergent then was that of Rothko. In these few years—from 1945 to 1950—artists as varied as Rothko, Gorky, Newman, de Kooning, Pollock, Still were engaged in vigorous debate with themselves and were willing to express their views publicly. They were not young hotheads. They had been schooled during the Depression years, and most were more than thirty-five years of age.

Rothko was forty-four in 1947 and living an intensely exciting moment. His first real success after the Art of This Century show had occurred in the summer of 1946 when the San Francisco Museum of Art held a large one-man exhibition of his work. He visited the West Coast and wrote back to his new dealer, Betty Parsons, that he had found real "adulation." Emboldened by his cordial reception, Rothko had thrown himself into his work with heightened enthusiasm and had redoubled his efforts to define, for himself first, and then for artistic friends, his proper aesthetic terrain. It was an unusually vivid moment in American art history. Rothko's participation in public forums can be seen in the context provided so richly by the two postwar publishing ventures, *Possibilities* and *The Tiger's Eye*, in both of which he offered carefully composed statements of his beliefs.

The Tiger's Eye appeared late in 1947. It was designed as a quarterly and for a time Barnett Newman was an associate editor, inviting his friends—many now represented by Betty Parsons—to contribute statements in a section he called The Ides of Art. Among them were Rothko, Baziotes, Hedda Sterne, Gottlieb, and Herbert Ferber. Newman's aesthetic position was, for a few years, parallel with Rothko's and Gottlieb's. His intellectual bent was more pronounced, however, and he was given to working out his ideas (and sometimes theirs) in written form. In 1945 he had written a manuscript in which he elaborated the themes Rothko and Gottlieb had broached in their *New York Times* letter. They are the themes that he was to introduce in 1947 in *The Tiger's Eye*, and from which he and his friends would most remarkably depart pictorially within just a few months. (The quick evolution of the Abstract Expressionists from semi-abstraction to abstraction is documented almost moment by moment in the publications.) Newman wrote in his unpublished 1945 manuscript:

> After more than two thousand years we have finally arrived at the tragic position of the Greek and we have achieved this Greek state of tragedy because we have at last ourselves invented a

new sense of all-pervading fate, a fate that is for the first time for modern man as real and as intimate as the Greeks' fate was to them. . . . Our tragedy is again a tragedy of action in the chaos that is society (it is interesting that this Greek idea is also a Hebraic concept), and no matter how heroic or innocent or moral our individual lives may be, this new fate hangs over us. We are living then through a Greek drama and each of us now stands like Oedipus and can by his acts or lack of action, in innocence kill his father.[30]

Within two years Newman, and everyone else who had brought "action into chaos" in their works, was pressing beyond the classical sources, though he still maintained that "the central issue of painting is the subject matter, what to paint." It was a sober, and, some would say later, rather humorless period. But the seriousness was beyond dispute. Even Arshile Gorky, who was somewhat disengaged from the group fervor, would write to his sister that "art must always remain earnest . . . Art must be serious, no sarcasm, no comedy. One does not laugh at a loved one."[31] Both publications would stress seriousness, earnestness, urgency, and would reflect the Abstract Expressionists' preoccupations as they worked their way through to total abstraction. Amid advertisements for books by Valéry, Kafka, Lorca, Sartre, and Stein, they scattered their thoughts concerning: the function of metamorphosis in art; moral despair and aesthetic nausea (Kierkegaard was in vogue); the meaning of primitive art; the role of mythology; the question of commentary and its relevance to painting, and the meanings brought home by the unprecedented carnage of the Second World War.

Rothko's two favorite literary sources—Shakespeare and Greek tragedy—were amply treated. In the second issue of *The Tiger's Eye*, Ariel's speech with its T. S. Eliot association, is reproduced. This speech not only addresses itself to mortality, as Rothko insisted painters must, but brings in the rich theme of metamorphosis, endemic to the mythic atmosphere he had already engendered in his work:

> Full fathom five thy father lies
>   Of his bones are coral made;
> Those are pearls that were his eyes:
>   Nothing of him that doth fade,

> But doth suffer a sea-change
> Into something rich and strange . . .

Shakespeare was also to be the source of the painterly passing beyond. When artists and writers hovered near the gateway to abstraction, they could find strange clues in Shakespeare. Rosenberg in *Possibilities* explores Hamlet's ambivalence and places his emphasis on the endlessly hermetic lines: "Seems, madam! nay it is; I know not seems." When Rothko would shortly after speak of bypassing obstacles, and when he would begin to purge his paintings of allegory, he could say, like Hamlet, "I have that within that passeth show." Hamlet's purification rings through Rothko's statement in the October 1949, issue of *The Tiger's Eye*. He sees clarity as the elimination of all obstacles between a painter and his idea, and between the idea and observer. Rosenberg cites Hamlet's speech:

> Yea, from the table of my memory
> I'll wipe away all trivial fond records,
> All saws of books, all forms, all pressures past,
> That youth and observation copies there;
> And thy commandment all alone shall live
> within the book and volume of my brain
> Unmixed with baser matter: yes, by heaven!

Rosenberg's comment is almost a program for those whom he would later, with Hamletian thrust, call action painters:

> Feeling himself unable to go forward, he must destroy the accumulations in which he has recognized himself and come to put his trust in the possibilities of the unknown.

This is just what Rothko does in his own meditation in *Possibilities* where he implies a leap into the unfamiliar. Calling his paintings "dramas," he says they begin "as an unknown adventure in an unknown space." His shapes, he says, "have no direct association with any particular visible experience, but in them one recognizes the principle and passion of organisms." Already the trivial fond records and forms and pressures of the past—the "ideas and plans that existed in the mind at the start"—are banished as the adventurer leaves the world in which they occur.

100

The move toward disembarrassment of the past reflects a consciousness of the bankruptcy—a word the painters often used—of past positions, and a dalliance with the notion of aesthetic despair. Motherwell in the issue of *The Tiger's Eye* in which Barnett Newman's interest in the classic notion of the sublime is explored, speaks of "getting rid of what is dead in human experience" and speculates:

> Perhaps painting becomes Sublime when the artist transcends his personal anguish, when he projects in the midst of a shrieking world an expression of living and its end that is silent and ordered . . . (#6, Dec. 15, 1948)

And Gottlieb in an earlier issue speaks of "a desperate attempt to escape from evil,"

> the times are out of joint, our obsessive, subterranean and pictographic images are the expression of the neurosis which is our reality. (#2)

As the painters tried to find their way in the labyrinth of sources newly examined, they often vacillated between a desire to retain a foothold in the intellectual world of Western civilization, and a desire to achieve the directness, the untramelled activity they imagined the primitive artists enjoyed. In the second number of *The Tiger's Eye*, there is a transcription of Kwakiutl shaman songs, as well as an important excerpt from Paul Valéry that supports the idea that an artist can act, rather than think, his creations:

> The idea of *making* is the first and the most human of ideas. To "explain" is always only to describe a manner of *making*: it is merely to remake by thought. The *Why* and the *How* which are simply expressions for what this idea requires, arise in every context, and insist on being satisfied at any price.

Valéry's importance to the "band of Myth Makers" should not be underestimated. They relished his unorthodox, sometimes aphoristic pronouncements, such as "for that which in us creates has no name; we have merely eliminated all men, *minus one*." Rosenberg was to draw extensively on Valéry's perceptions of the creative process as he worked to give definition to the new activities amongst painters. Baziotes in *Possibilities* chooses a text from *Variété* on

"The Silence of Painters." Those who were thrashing out the issues repeatedly reverted to the classical myths, but little by little their endeavor was transformed into a quest for an absolute that the classical world would not support. Thoughts raced ahead but the paintings reproduced in the magazine often alluded to the ancients. In March 1948 there was de Chirico's "Hector and Andromache," Stamos's "Sacrifice of Cronos," Masson's "Bull Head, Twisted Figures," Gottlieb's "Pictograph," Rothko's "Sacrificial Moment," Gorky's "Agony," and even de Kooning, the least mythic in spirit, has a painting called "Orestes." Newman, beginning his polemic for an American art that distinguished itself from the highly civilized European (reflecting the ambivalence toward Europe, springing from the 1930s and becoming something not so remote from the hated nationalism of that epoch) writes in the same issue that the American in comparison to the European is "like a barbarian"

> He does not have the super-fine sensibility toward the object that dominates European feeling. He does not even have the object. This is, then, our opportunity, free of the ancient paraphernalia, to come closer to the sources of tragic emotion. Shall we not, as artists, search out the new objects for its image?

Shortly after, in the symposium on the sublime in art,[32] he is speaking of "man's natural desire in the arts to express his relation to the Absolute" and again appoints himself spokesman for the Americans who reject "concern with beauty and where to find it"

> We are reasserting man's natural desire for the exalted, for a concern with our relationship to absolute emotions. We do not need the obsolete props of an outmoded and antiquated legend. . . . We are freeing ourselves of the impediments of memory, association, nostalgia, legend, myth, or what have you, that have been the devices of Western European painting.

Along with the new tone of rebelliousness, and a sense of security that American painters had never known vis-à-vis their European fathers, there was a concomitant alertness to everything new that appeared in postwar Europe. The big thinkers still seemed to be anchored there, and the publications the artists wrote in and read, were quick to pick up the new currents abroad. *The Tiger's Eye* respected French intellectuals and printed translations of work by Georges Bataille, Genêt, René Char, and the Surrealists' favorite, the Comte de Lautréamont. In addition, they were attentive to the immediate postwar painters, some of whom were exhibited very soon after the war in New York, among them Fautrier, Matthieu, Michaux, Ubac, and Wols.

Rothko's attitudes were subtly changing as can be seen in his last statement made for *The Tiger's Eye*, October 1949. "The progression of a painter's work, as it travels in time from point to point, will be toward clarity: toward the elimination of all obstacles between the painter and the idea, and between the idea and the observer." Examples of such obstacles he gives as memory, history, or geometry, "swamps of generalization from which one might pull out parodies of ideas (which are ghosts) but never an idea in itself." "The idea in itself," was to become his leading idea. It is very much dependent on the original impulse that had suggested to him that music could be in some way a source of his own means. He wound up his statement confidently: "To achieve this clarity is, inevitably, to be understood." Rothko uses this "understood" in the broadest sense, assuming that the observer is moved, he knows not how, and that through the purity (clarity) of the "idea" comes another kind of knowledge, not dependent on explanations. The brevity of this final statement of the 1940s speaks of his growing conviction that words, words, words, as Hamlet declared, could not substitute for a painter's making his paintings. Both he and Clyfford Still had begun to make an issue of the uselessness of words, and the triviality of art criticism. Still wrote letters to Betty Parsons, and to Rothko, decrying the "scribblers" and insisting on their irrelevance in the face of the painters' great decisions to become totally independent. The abstraction "freedom" began to assume new significance as these two painter anarchists were preparing their assault on the organized art world. Their defiance of what they saw as the vulgarity of the art world is hinted at throughout the late 1940s, but becomes more arrogant in the 1950s. In *The Tiger's Eye* (#2, 1947) Rothko had already suggested the extravagant ethical basis on which they would proceed when he stated that "a picture lives by companionship, expanding and quickening in the eyes of the sensitive observer. It dies by the same token. It is therefore a risky and unfeeling act to send it out into the world. How often it must be permanently impaired by the eyes of the vulgar and the cruelty of the impotent who would extend their affliction universally!" Rothko could easily see that American philistinism was not extinguished, despite the new interest in the band of mythmakers. A touch of the old reformer's ambition was left, and this was kindled by Still's hyperbolic personality. When Douglas MacAgy invited Rothko—at Still's suggestion probably—to come to teach a summer semester at the newly reorganized California School of Fine Arts, Rothko offered not only studio instruction but a series of lectures as well.

The atmosphere at the school was bracing for Rothko. The students admired him, treated him as a master. They knew his work from the preceding year's retrospective, and some, such as Richard Diebenkorn, had spent hours studying "Slow Swirl by the Edge of the Sea" that had remained on view in the museum. Clyfford Still, who had a way of inspiring his students with fierce loyalty, passed them along to Rothko. Even those, who, like Diebenkorn, had found Still's extravagant polemics distasteful, were willing to submit to the more meditative approach Rothko offered. Many of the students were young men who had returned to school on the GI Bill. They were only too eager to believe that there could be a totally new expression in painting. Most were quite willing to hear Still denounce the European forebears and exhort them to follow their intuitions. Under MacAgy's inspired leadership, the school was, as he said, a "wide open situation." Anything could, and was meant to happen there. Still hectored his students and insisted on the ethic in which the painter takes full responsibility for what he does, and protects it from the hostile world. Rothko agreed, and was probably stimulated by the atmosphere in the school—a slightly hysterical atmosphere in which students and teachers, as MacAgy put it, knew they were making history. Rothko was considered a "very inspiring teacher," whom MacAgy described as smoking endless cigarettes, "the curling smoke almost a symbol of how he talked: very elusive talk." MacAgy's attention to Rothko's presence can be gauged in an article—the first serious article on Rothko in an art journal—published in *The Magazine of Art* (January 1949). It can be assumed that MacAgy's interest in Rothko's lectures, coupled with his early recognition of Rothko's distinction as a painter of new subject matter, informed his article. From the beginning he establishes Rothko's effort to give broadly philosophic meaning to his enterprise. MacAgy starts by stating that if we subscribe to the notion of painting as a symbolic act, then we can understand what Rothko means when he says that a painter commits himself by the nature of the space he uses. This was, undoubtedly, one of the principles Rothko had enunciated in his lectures. The bold departure from the quantified space of the Renaissance, stressed by the Surrealists, was of great importance to him. MacAgy sees Rothko moving in a space characterized by a continuous fading in and out, existing only as a relationship of things. Rothko himself, the author points out, "refers to the relations between objects and their environment as *dramatis personae* of his paintings" and he speaks of "purposeful ambiguity." MacAgy might be quoting Rothko when he states: "Pictures are episodes of transformation which engage the artist's interests in dramatic action." MacAgy understood (as few others did) that the "theatre of Rothko's imagination" displays the basic assumptions from which philosophies are formed. Although the work is visual, presented through sight, the experiences, MacAgy says, transcend the limits imposed by visible particularities.

MacAgy was referring largely to works produced between 1944 and late 1948, some of which had been exhibited in New York in March 1947 at the Betty Parsons Gallery and again in March 1948. The reviewers were not so prescient as MacAgy. Sam Hunter writing in the *New York Times* (March 14, 1948) accurately described Rothko's art as "an art solely of transitions without beginning, middle or end" but, unlike MacAgy, saw it as "an impasse of empty formlessness." The paintings of 1947-48 were, in fact, ambiguous, showing Rothko feeling his way toward an expression that would directly, without the interference of specific shape, suggest the numinous floating world into which he stepped once his obstacles were left behind. He scraped and thinned his colors. Their edges bled. A form with the most tenuous of edges could slide into another, while atmosphere from one moment to the next could be weighted or lightened. A vocabulary of starts and stops, almost like

21. Untitled   1949

musical notation, is contained here, with short strokes sometimes interrupting longer ones, and rounded forms juxtaposed with long, strand-like shapes, or vertical pole-like shapes, none of which seemed very stable, and all of which were subject to minute transformations. After his show in 1945 Rothko worked often on large watercolor paper with an exceedingly toothy grain. By dragging a dry brush on slightly dampened surfaces he had discovered many ways to suggest mirrored forms, and forms that seemed to emanate their own light. Some of these watercolors were composed with three levels, each illumined differently and each furnished with ambiguous signs relating to the others. No doubt these freely explored motifs, and particularly the surprising effects of light, encouraged the radical changes in his works in oil on canvas. These more abstract works were still "anecdotes of the spirit" and the drama that Rothko had said inheres in a Rembrandt portrait could often be read in moody tones that masked lightly sketched undertones. Rothko had almost worked his way through to the point that he had envisioned in 1943 where "the whole of man's experience becomes his model and all of art is the portrait of an idea." When Harold Rosenberg came to summarize the works of the abstract expressionists, referring specifically to Rothko and Newman, he wrote

> In sum, what the new American artist sought was not a richer or more contemporary fiction (like the Surrealists), but the formal sign language of the inner kingdom—equivalents in paint of a flash, no matter how transitory, or what had been known throughout the centuries as spiritual enlightenment.[33]

The "flash," toward which Rothko had instinctively directed himself, was about to occur.

Private enlightenment was not openly avowed, however, not even to himself. Rothko was caught up in the missionary fervor that motivated many painters during the late 1940s. He still attempted to carry out the public education that had once been the function of the cultural committee of the Federation. It was commonly believed among painters that unless the public—and that included the critics—was taught to see what it saw, it would never see it. The painters were not wrong. There were still very few writers con-

22. Untitled   c.1946

cerned with their work, and, apart from Greenberg and Rosenberg, none from the hallowed literary community. The attitude of most of the literati was one of patronizing wonder. They were convinced that painters were somehow not capable of dealing with thoughts and ideas, and that the burgeoning aesthetic was remote from their own preoccupations at that moment. This persistent secret contempt for the handworker has never completely abated in the United States. When one of *Partisan Review's* former editors, William Barrett, wrote his memoirs he alluded fondly but patronizingly to the Abstract Expressionists, especially Willem de Kooning and Franz Kline.[34] He recalls giving a lecture on Existentialism at the Artists' Club but noted that "ideas, abstract ideas have a way of bouncing off the minds of artists at curious angles." Barrett shrunk from giving a second lecture: "I remember remarking to one of them that perhaps they had enough of ideas and should stick to their art." He doesn't hesitate to add that the New York School history was to be "an indecent traffic with ideas, in the course of which it is really remarkable that some good painting managed to get done."

In the face of such stubborn prejudice, Rothko and other artists were at once enraged and exasperated. The Club itself was an outgrowth of their frustration in the general culture. It grew out of the discussions Rothko, Still, and MacAgy commenced in 1947 about a school for young painters that would be conducted in an entirely unorthodox manner and would focus on the real problems and ideas the postwar period had spawned. In the fall of 1948, Rothko, Gottlieb, Baziotes, Motherwell, and David Hare founded the school, called, at Newman's suggestion, Subjects of the Artist. A lecture program was initiated, open to the public. For Rothko the stress of the school proved too much (probably the stress of internecine wrangling) and in midwinter he wrote to Still that he was on the eve of a nervous breakdown and was withdrawing from the school, which itself folded in the spring. The lectures continued, sponsored by another group of painters from New York University, and eventually the Friday nights came to be consecrated as the informal entity known as The Club. To this often disorderly and always lively gathering came an assortment of New York intellectuals, curious about *la vie Bohème* personified by the hard-drinking, sometimes rowdy artists. Only a few, however, understood that painters were indeed ploughing in the same fields and using the same "ideas" that they were. The most serious inquiry was undertaken by Clement Greenberg in both *The Nation* and *Partisan Review,* where he wrote regularly about the group. He piqued the curiosity of his timid literary confreres when he wrote of the poverty of the painters, their fifth-floor cold water studios, their "neurosis of alienation," and the increasing sense of wild abandon he felt in the work, particularly of Pollock. Greenberg's doughty prose, and Harold Rosenberg's witty exposition of the general moral atmosphere among these newly invigorated artists, did help to attract an ever-widening audience. But Rothko sensed, perhaps correctly, that this audience, while friendly, was still unable to decipher the new language and was indiscriminate in its taste. Little by little he relinquished his public activities. He turned up rarely at The Club, but sometimes appeared in the favored bars to exchange views with fellow painters. Most of his conversations, however, took place in the homes of the artists with whom he showed at the Betty Parsons Gallery, or in the studios of a few of his more intellectually inclined colleagues.

7

By 1950 Rothko had talked himself, thought himself, into his position and then fallen relatively silent. He had struggled with prevailing ideas, often contradictory: on the one hand, the strong tendency among American intellectuals to endorse practical positivism and on the other, his own instinctive drive toward what Rosenberg had called "the flash." Through it all, he had managed to preserve the original insights drawn from Nietzsche. He still saw his art as an effort to express his sense of tragedy. He had also begun to decipher for himself the language, the nature of the expression, that had been born long before the First World War, and that later, in the hands of Malevich, Mondrian, and a few other absolutists, constituted an authentic way of pictorial thought. Henceforth his arguments with himself would be couched in the esoteric, almost inaudible dialect that had finally revealed itself to him. Perhaps he had taken to heart Nicolas Calas's counsel in the December 1948 issue of *The Tigers' Eye* that "One should listen to the stillness of painting with the awe with which one harkens the silence of deserts and glaciers."

Rothko was no more at ease with life than he had ever been. The "everyday" was still a difficult realm. He was more avid than ever to know, fully, the satisfaction of release from necessity and to "hark to the suspiration/the uninterrupted news that grows out of silence."[35] His new preoccupation with silence had its metaphysical dimension, much as he fought shy of it. Every once in a while, though, a secret yearning shone through his spoken statements, as when he had admitted ironically in *Possibilities* that the primitive artist lived in a more "practical" society where "the urgency for transcendent experience was understood, and given official status."[36] Implicit in his statement is an attack on materialism. He speaks of "the finite associations with which our society increasingly enshrouds every aspect of our environment." He laments solitude. Yet, it was silence and solitude which would lead him to the transcendent experience he implied every society must know in order to be healthy. It was a silence that others were beginning to explore. In the same issue of *Possibilities* in which Rothko was speaking, Baziotes was presenting a text by Paul Valéry called "The Silence of Painters," and Paul Goodman was writing on "primary silence" and Nietzsche.

These silences, paradoxically invoked in a time of much noisy discussion, had their origins in Europe and were transmitted in many untraceable ways to American artists. Rothko was susceptible on the deepest level—he had long been familiar with the "uninterrupted news" while at work. Since late 1945 he had managed to live in the vicinity of the Museum of Modern Art and could drop in whenever the spirit moved him. It moved him frequently in the late 1940s and early 1950s. There, in an atmosphere considerably more calm than it is today, he could study the permanent collection, in which certain works had their permanent place. Alfred Barr and René d'Harnoncourt had installed the permanent collection sensitively. The silence of certain modern precursors was reverently respected. As if to stress the point, Redon's painting in the high symbolist manner, in which a thinly painted, ethereal personage holds a silencing finger to its lips, remained for years in the first room of the permanent collection. Its relation to Matisse would not be overlooked.

It was to Matisse that Rothko gravitated with renewed interest. In his old concourse with Milton Avery the polite respect awarded Matisse was apparent. But in his new frame of mind, Rothko could respond to Matisse's radicalism without reserve. Years later he would call upon his wife as witness, recalling how he had spent hours and hours and hours before "The Red Studio" once it was permanently installed in 1949. It was of crucial importance to him he later thought. Having come so far himself in conjuring an atmosphere uniquely other than the "everyday," Rothko could confirm many of his intuitions through the study of "The Red Studio." The paradox implicit in Matisse's firm statement was poignantly alive to Rothko. Matisse presents what can be read as a room, a studio, with a number of identifiable objects—brushes, dishes, jugs, chairs, easels, stools, plants, sculptures, paintings, and a clock. There are things there, things that bespeak a human presence, things that are used in everyday life, things that Rothko admired Avery for being able to incorporate in his life's work. And yet, these "things," with which Rothko himself was so ill at ease, were not really there—as Matisse was careful to emphasize—*as* things. Their barest outlines were merely absences that became outlines, as Matisse left the white of the canvas around them. Their presence was to be a virtual presence in an overwhelming atmosphere that Matisse announced as whole; as, in fact, a personal cosmos. The centered clock gives the circular time and the objects rotate within a whole, irradiated with an inner light that was always Matisse's goal. "The destination," as he said so often, "is always the same." If he worked with thinly layered paint in impeccably modulated reds, it was to achieve the dreamed-of unity that could be found in the light of the mind.

This disembodied light had visited Matisse as a youth and he never renounced it. It was part of the tradition that he claimed and sustained throughout his life—the tradition so brilliantly expounded in the French Symbolist movement. Throughout his painting life Matisse reflected the principles implicit in the Symbolist attitude toward reality. Once he had passed his psychological condition of being a student, he had thought ceaselessly of the meanings inherent in his painting and had found them consistent with the meanings adduced by the Symbolists. Art exists not to represent the world as it is lived day by day, but to create a parallel world—a "condensation of sensations." He would, he said, sacrifice charm in order to seize the essential character of what he is painting. "Underneath this succession of moments which constitutes the superficial existence of beings and things, and which is continually modifying and transforming them, one can search for a truer, more essential character." For him, "all is in the conception." Therefore, he must have a clear vision of the whole from the beginning. All of this Matisse stated by 1908.[37] Toward the end of his life, he was still speaking of his "internal vision": "Thus a work of art is the climax of long work of preparation. The artist takes from his surroundings everything that can nourish his internal vision, either directly, when the object he is drawing is to appear in his composition, or by analogy."[38] This analogy, to which Matisse often refers in his interviews and statements, is the spirit of analogy articulated by the poet Mallarmé, whose poems Matisse read early in the morning "as one breathes a deep breath of fresh air." Mallarmé had told himself as a youth that he must purify language and use it so that it could "describe not the object itself but the effect it produces." He spoke of "a spiritual theater" and an "inner stage" where absences (equivalent to the whites Matisse employed) evoked his inner drama. The objects in his poems were to be assembled in the receiver's imagination symbolically. "Things exist, we don't have to create them, we simply have to see their relationships," Mallarmé wrote, "Our eternal and only problem is to seize relationships and intervals, however few or multiple." On his part, Matisse observed: "I don't paint things, I only paint differences between things."

Matisse had not been insensible to the long tradition going back to Baudelaire and Delacroix in which the "musicality" of painting was perceived as more significant than the individual motifs within a painting. He had studied the works of their natural heirs—Gauguin and Cézanne—assiduously. Gauguin's words, well known to painters emerging in the 20th century, clearly echoed Mallarmé's;

> My simple object, which I take from daily life or from nature, is merely pretext, which helps me by means of a definite arrangement of lines and colors to create symphonies or harmonies. They have no counterparts at all in reality, in the vulgar sense of that word; they do not give direct expression to any idea, their only purpose being to stimulate the imagination—just as music does without the aid of ideas or pictures—simply by that mysterious affinity which exists between certain arrangements of colors and lines in our minds.[39]

It was to this view, epitomized by Mallarmé, that Matisse, and, through Matisse, Rothko, gravitated. Until Rothko undertook his most taxing adventure into the unknown that he had said was the artist's province—the mural series that dominated his last phase—he was playing out the lessons implicit in the symbolist tradition. He had said he wanted to raise painting to the level and poignancy of poetry and music. The Symbolists had begun the project, and he was to extend it beyond their most extravagant dreams.

His means was painting, and only painting, but revery was its source. Releasing the "slow swirl" of revery was as familiar an occupation to him as walking or smoking a pipe, as he had in his youth. He could entertain in his reveries the possibility of miracle,

> Pictures must be miraculous: the instant one is completed the intimacy between the creation and the creator is ended. He is an outsider. The picture must be for him, as for anyone experiencing it later, a revelation, an unexpected and unprecedented resolution of an eternally familiar need.[40]

Like other painters of his generation, he searched the horizons of many ages, including his own, in order to find miraculous expressions of eternally familiar needs. Increasingly the silences of certain epochs and artists inspired his reveries. He must have stood many times before the Pompeian frescoes installed at the Metropolitan Museum, where he frequently went to study Greek and Roman antiquities. The calm figures of Pompeian matrons and the presences of the gods, repose in a hushed place in a perfect harmony of parts. The reds, from pomegranate to crimson, are genially faded, urging upon the viewer still more the remoteness of the dream-ridden characters. The Pompeian way was largely theatrical. There is always the breath of theater in the way the painters provide a shallow stage and an appropriate backdrop. What is occurring here? Not even the owners of the villa knew exactly. Years later, when Rothko finally saw Pompeii itself, he met with familiar emotions. The House of Mysteries, with its grave and hermetic theatrical rites, meets his Nietzschean requirements. Rothko recognized himself in Pompeii. His recognition of himself and his own purposes occurred long before he was recognized, or rather, understood by those outside his inner circle of friends. Unlike Pollock and de Kooning, Rothko, as Motherwell has pointed out, suffered from not

23. Rothko at Betty Parsons Gallery, New York, 1949

having a strong advocate. Each time he exposed himself to the appraisals of others he was met with either hostile incomprehension or reticent, mumbled praise, with only very rare exceptions. He tried to brazen it out. In *Possibilities* he had consoled himself with the thought that in the absence of an embracing approving community the artist gains in freedom. "Both the sense of community and of security depend on the familiar," he had written. "Free of them, transcendental experiences become possible." All the same, he longed for others to recognize in his work the resolution of "an eternally familiar need." That was to remain in abeyance for a long time to come. In the late 1940s and early 1950s, his exhibitions at the Betty Parsons Gallery met mostly with such bewildered responses as Margaret Breuning's (*Art Digest*, April 15, 1949), who remarks that "schoolroom Latin comes to mind in viewing the exhibition of Mark Rothko's paintings at Betty Parsons Gallery, for Virgil's *disjecta membra* exactly sums up the impression of these amorphous works." Thomas Hess in *Art News* was much more sympathetic, although he too limited his remarks as though he were awaiting further enlightenment. He notes Rothko's "wild contrasts of color" and that "under this extremely emotional level is a strength of composition which is almost Oriental." It would be several years before Hess would write with strong conviction about Rothko's paintings. Even then, if the literature on Rothko during the 1950s is surveyed, it will be seen that he attracted far fewer serious articles than many of his colleagues in the New York School, and often even his most admiring commentators could not conceal questions and doubts that were rarely raised in the case of the others. No one with any knowledge of New York School painting would have doubted his importance, or underestimated his radical gesture. But something was withheld; some intangible sign of understanding for which Rothko was longing. The unreserved response, and the more searching approach to his work would come, eventually, from Europe and England.

It is not surprising. Europeans were backed by a tradition to which Rothko's aspirations could easily be assimilated. They quite naturally apprehended the expression in Rothko since they themselves had been working in parallel directions, in which a convergence of previous expressions and theories flowered after the Second World War. Nietzsche had been resurrected in France first by Gide and later to serve the Existentialists. In France poets, philosophers, and painters were undertaking an intense inquiry into the nature of space and how man inhabits it, and the nature of the human drama and how man interprets it. Sartre and Merleau-Ponty were fashioning their fresh theories of perception, in which they often called upon artist witnesses, and Camus was questioning the nature of creativity. In such an intellectual climate, those susceptible to Rothko's restatement of the old problem of Mallarmé—the absence, the void with its metaphysical echoes—were well equipped. Rothko came, bringing what Rilke called the "suspiration that grows out of silence," and it was understood by those who were listening to the same all but inaudible sounds.

If, as Rothko thought, a picture lives by companionship, there were to be eyes that were prepared to be companionable in Europe. The kind of pictorial thinking Rothko's abstractions represented after 1950 found immediate recognition in Europe in the later 1950s. To some degree, Rothko's thought was itself tempered by the cultural news arriving from Europe. In the early 1950s he saw Motherwell and Philip Guston often—both, artists intensely interested in the way the philosophies being formulated in postwar Europe could enhance their own. Guston was an avid reader and questioner, as was Motherwell. The sources frequently cited by the French writers—Nietzsche, Kierkegaard, Kafka, Dostoyevsky—were already familiar to them, important to them. As these painters stepped out gingerly into the unknown, they looked behind them and they looked forward, scanning the intellectual horizons seeking moral support. It was slow in coming. But they could find confirmation of their daring departures in painting in the respectful attention rendered painting by the most fertile French minds of the postwar era. What they were doing, in the eternal reciprocal round of stylistic choices, could be interpreted comfortably by the European mind that had already moved beyond formalism. Even the most intelligent commentators in America, such as Clement Greenberg, had not yet acknowledged that the kind of "transcendental" experience Rothko posited was possible in painting. Rothko's most serious apologist, and the one he most respected in America, Robert Gold-water, was also loath to admit a metaphysical dimension to pictorial thought, and insisted in the 1960s that such speculation was "literary fancy."

If there is a context within which Rothko's work may be considered as part of the modern history of thought, it would be found in the peopled silences invoked by a host of French writers from 1945 on—not only philosophers but poets, playwrights, and critics. It would be found in the continuity evident between late 19th-century Symbolism and mid-20th-century Existentialism. It would be found in the resurrection of Nietzsche, especially in France where Gide had first invoked him to support his fin de siècle rebellion against rationalism, and where Sartre and Blanchot would later find important matter for their definition of the modern spirit. Rothko had arrived at his own Nietzschean vision when he spoke of the impossibility for the solitary figure to raise its limbs in a single gesture that might indicate its concern with the fact of mortality (1947) and when he said "I belong to a generation in which every artist studied the human figure. It was with the utmost reluctance that I found this figure could not serve my purposes. A time came when none of us could use the figure without mutilating it. If I couldn't find ways of dealing with nature without mutilating it, I felt I had to find other ways to deal with human values" (1958). His personal re-valuation of values was to take the form of a stripping down, a reduction to the pregnant silences already envisioned by the Symbolists. Nietzsche thought that "the will loves better to will noth-ingness than not to will." When Nietzsche began his own surpassing effort to separate and rub clean each of his thoughts and words, and to rid himself of meaningless bulk, he took into account the silences from which could arise the purest of thoughts, much as had his contemporary Mallarmé, and also Rodin, who designated the fragment an emblem of a whole. "The aphorism," Nietzsche wrote, "is a form of eternity; my ambition is to say in ten phrases what another says in a book—does not say in a book." This fanatic thirst for essence would be transmitted in many forms to 20th-century artists. It is the very premise of Existentialism, as Sartre insisted. Subjectivity, he said, must always be the starting point since, in his view, existence precedes essence. The one thing man cannot do is to transcend subjectivity. His whole life is a process moving toward essence, which, in effect, is a kind of transcendence. Despite man's solitary situation, he only becomes aware of himself (becomes his essence) through the *cogito* that also perceives all others, and perceives them as a condition of his own existence. Here is the world of "intersubjectivity" which Rothko anticipated when he spoke repeatedly of the "others" who beheld his pictures, and their "human needs." Rothko was thoroughly prepared, emotionally and intellectually, to consent to the Sartrean vision of art as functioning in the same way as ethics "in that we have creation and invention in both cases. We cannot decide, *a priori* what there is to be done." What Sartre wrote in "Existentialism Is a Humanism," a rather simplified 1945 lecture published as *Existentialism* in New York in 1947, was perfectly consistent with Nietzsche and what Rothko had assumed from Nietzsche:

> Man is constantly outside of himself; in projecting himself, in losing himself outside of himself, he makes for man's existing; and on the other hand, it is by pursuing transcendent goals that he is able to exist; man, being this state of passing-beyond, and seizing upon things only as they bear upon this passing beyond.

Passing beyond—*outrepasser*—was a key idea found everywhere in the writings meant to interpret the cultural moment that arrived after the Second World War. Beckett, Malraux, Ponge, Camus, Sartre, Blanchot, Merleau-Ponty, Bachelard: all were preoccupied with the word even if, as several of them underlined, it was quite possible that such passing beyond was not possible. Whatever it might mean to individual sensibilities, it had a broad general meaning at the time that implied that all is in process at all times, and a part of such process is to transcend the immediate givens of a situation and to understand, as Nietzsche had said, that there are no facts in themselves. Everything is perceived in a situation. Reciprocities are infinite. The readiness of the French thinkers to admit painters into the thinking process was epitomized by Maurice Merleau-Ponty whose writings are permeated with his profound experiences with painting. Merleau-Ponty more than any other postwar writer was able to encompass the acts of the painter in an overall philosophy in

1. *Interior* c.1932

2. *Birth of Cephalopods*   1944

3. *Slow Swirl by the Edge of the Sea*   1944

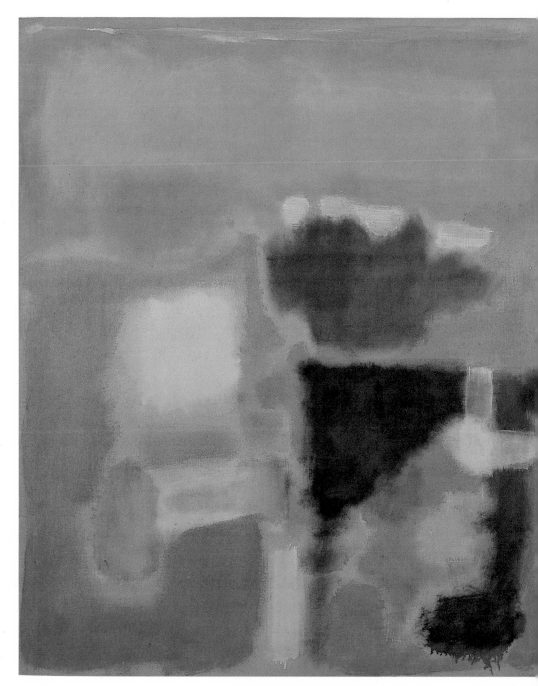

4. *Number 18* 1948-49

5. *Number 11*   1949

6. *Number 61* 1953

7. *Number 18*   1951

8. *Blue over Orange* 1956

9. The Rothko Room at The Tate Gallery, London

10. Untitled   1958 Full-size study for the Harvard Murals

11. *Number 117*   1961

12. Untitled 1969

13. Study for Seagram Murals   c.1958

14. *Number 17* 1947

15. *Greyed Olive Green, Red on Maroon* 1961

16. *Triptych*   1965-67, The Rothko Chapel, Houston, Texas

which the language of the painter and the language of the writer were assumed to have common traits. He returned to the great modern tradition born in the genius of Mallarmé and Cézanne in order to understand contemporary art with its implacable will to reduction and aphoristic trenchancy. What Mallarmé understood of language (and his drive to pass beyond it) was its infinite plasticity, its extension into spaces and silences that could not be contained by syntax. He had spoken of two languages: the one he called practical, and the other echoed in the self-mirroring words of his poems. Merleau-Ponty follows Mallarmé:

> Speech in the sense of empirical language—that is, the opportune recollection of a pre-established sign—is not speech in respect to an authentic language. It is, as Mallarmé said, the worn coin placed silently in my hand. True speech, on the contrary—speech which signifies, which finally renders *l'absent de tous les bouquets* present and frees the meaning captive in the thing—is only silence in respect to empirical usage, for it does not go so far as to become a common name.

The passage Merleau-Ponty refers to is in Mallarmé's "Crisis of Poetry":

> If a poem is to be pure, the poet's voice must be stilled and the initiative taken by the words themselves, which will be set in motion as they meet unequally in collision . . . I say: "a flower!" then from that forgetfulness to which my voice consigns all floral form, something different from the usual calyces arises, something all music, essence, and softness: the flower that is absent from all bouquets . . .[43]

The painters are as familiar with this "authentic" language of absence as the poets. There is a tacit language, Merleau-Ponty insists. Both poets and painters speak it and

> We must consider speech before it is spoken, the background of silence without which it would say nothing.

What could be more evident in Rothko's thinking, through his painting? The spaces discerned throughout the poem and the painting are spaces of a modern point of view of existence, spaces that are not outside or inside, but are endemic to all being; spaces, as

Merleau-Ponty says, reckoned only as starting from the individual as the zero point or degree zero of spatiality. On Space:

> I do not see it according to its exterior envelope. I live in it. After all, the world is all around me, not in front of me . . . The question is to make space and light, which are *there,* speak to us.[44]

This he felt Cézanne had done. (Matisse perhaps drew his inferences from Cézanne, for he had stated, in his work and in his explanations of his work, that the space is all around him, "the wall around the window does not create two worlds." And, "Often I put myself into my picture and I am conscious of what exists behind me.") Rothko had arrived through his process at a similar conviction concerning the nature of space, and in 1951 explained that he painted large pictures

> precisely because I want to be very intimate and human. To paint a small picture is to place yourself outside your experience, to look upon an experience as a stereoptic view or with a reducing glass. However you paint the larger picture, you are in it . . .[45]

What the French phenomenologists kept referring to as the "lived" experience was to become Rothko's language, his tacit language, and his pictorial language. As Merleau-Ponty understood (admonishing his readers that there is so much to be learned about thinking from painters), "It makes no difference if he does not paint from 'nature'; he paints, in any case, because he has seen, because the world has at least once emblazoned in him the cipher of the visible."[46]

Painting was Rothko's means, his only means to convey what he called human values that were experienced as a passing beyond; as what older civilizations had called the states of the soul. In order to bring painting to the degree of emotion he had known in music, or in poetry, he had to forgo the naming of things that Mallarmé had consigned to practical language. Henceforth he would attempt to name only his emotions in the grandest, most ardent ways, bearing in mind always the ambivalent situation of modern man. The problem of modern painting, as Rothko had understood, was how to transcend consciousness of self, or, as Merleau-Ponty put it

Modern painting presents a problem completely different from that of the return to the individual: the problem of knowing how one can communicate without the help of a pre-established Nature which all men's senses open upon, the problem of knowing how we are grafted to the universal by that which is most our own.[47]

Rothko in the 1950s was prepared to go beyond Symbolism to sense the silences behind and beneath his every gesture on the canvas. The forms in his paintings would still be like "actors," but now they acted in a different drama in which anecdote disappears into light. The paint would speak, as words would speak in Mallarmé, of other things. He would believe, with Redon, that *"la matière a son genie."* The "lived" experience would be transmitted as directly as possible through paint and he would be, as he early dreamed, the *chef d'orchestre*, and the composer of a great opera like *The Magic Flute:* "Every painting I do now is different, as if I were writing *The Magic Flute*—one day Sarastro, one day Pamina, and so on."[48] In going beyond the symbolism that underscored his work of the late 1940s Rothko ceded to his impatience with practical language, and with the encrusted usages that impaired his vision. He was hearing another sound in the world that pierced to a center. "I know not seems," Hamlet had said.

In 1950, a seasoned painter-turned-art historian, William Seitz, embarked on a thesis that would attempt to explore the foundations of the movement called Abstract Expressionism through examining carefully the works of several artists. He was well liked by his subjects, one of whom was Rothko. Seitz spent many hours in studios, looking and talking. He frequented the artists' cafés and bars and The Club. He compared what was said in one studio with what was said in another. His deep intelligence sifted responses and shaped a picture of the movement that has never been surpassed. Naturally, he had his preferences, and if his work is read closely, Rothko seemed to him to be the most extraordinary of painters (confirmed by Seitz's own painting at the time). Seitz's conversations with Rothko were intense, and although not always directly quoted, inform his view of certain aspects of the movement. In their light, he considers the problem, for instance, of the way American painters approached the notion of the transcendental. Noting their commitment to process, and to the importance of matter itself, Seitz carefully defined the way these artists used the word transcendental to "indicate values, which, though subjective, are not merely personal. They are ideal or spiritual, but still immanent in sensory and psychic experience." He understood Rothko above all to be concerned with such values, and that in his instinctive drive toward an absolute, Rothko was struggling to elicit meanings unmediated by discursive language, or its formal equivalent in painting. Seitz called upon Sartre (and it is probable that he had discussed the passage with Rothko) to help him elucidate:

> Tintoretto did not choose that yellow rift in the sky above Golgotha to *signify* anguish or *provoke* it. It is anguish and yellow sky at the same time. Not sky of anguish or anguished sky; it is an anguish become a thing, an anguish which has turned into yellow rift of sky, and which thereby is submerged and impasted by the proper qualities of things, by their impermeability, and that infinity of relations which they maintain with other things.[49]

Seitz remarks that "Rothko, too, describes his areas as 'things' and like Sartre, emphasizes the 'immanence' of their meaning. Meaning is immanent in the form or, in Sartre's phrase, 'trembles about it like a heat mist'; it *is* color or sound." The passage in Sartre arrested the attention of Merleau-Ponty as well, and he commented that[50] "if we set ourselves to living in the painting, the meaning is much more than a 'heat mist' at the surface of the canvas, since it is capable of demanding *that* color and *that* object in preference to all others, and since it commands the arrangement of a painting just as imperiously as a syntax or a logic. For not all the painting is in those little anguishes or local joys with which it is sown: they are only the components of a total meaning which is less moving, more *legible*, and more enduring." Rothko would have liked this qualification as he was always wary of ecstatic interpretation, although what he sought for himself in his painting was precisely ex-stasis.

All of the clarifying, both in his reveries about painting and in the process of painting, that Rothko undertook around 1949 could not shake his fundamental doubt—a quality inherent in his personality. But it was a new kind of doubt, one focused entirely within the process of painting itself, the kind of doubt Matisse knew when so surprised was he at what had arrived on his canvas that he told a visitor that not he but the postman had painted "The Pink Onions." It was, so to speak, a positive doubt, one that could be turned into energy. Rothko's love of the prefix "trans"—above, over, through, beyond—grew imperious. How many times he had spoken of transformations, transmigrations, transfigurations, transactions, translation!

The "actors" or "things" on the face of his canvas move into place with increasing formality, as though the symmetries of ancient dramas had overcome Rothko's dreams of immemorial creation myths. They float, as in "Number 18" of 1948-49, but in an equilibrium established by near-horizontal balances, and in the apparitional bar—Rothko's familiar reflex to invoke other worlds. The *aura*, never absent from now on, makes its appearance. Boldly he announces his ambiguities. In what he called his "multiform" paintings, reds are moving both inward and to the surface without visible boundaries, and they are nimbused with a pinkish glow, or sent floating behind a rough rectangle of blue. Shapes that are deliberately divested of boundaries, or are pale specters of rectangles, are posited in order to speak of verticality, or of the masking of space by means of light. Memory will speak. The experiences Rothko has known in the act of painting and in his moving from point to point, as he put it, are given their equivalents in reductions to essences. They are remembered in paint. They are spread lightly on the canvas surface, and then overspread. They glimmer or joyously radiate from behind. A manifest preoccupation with what can only be called the apparitional is everywhere (or everywhere and nowhere as Pascal might have said) and recalls the Symbolists' obsessions with murmuring silences. In "Number 11" of 1949, Rothko has recourse to the mirroring symmetries they, the Symbolists, expounded. The bars of whitened blues, reds and pinks that travel upwards on the vertical format are like bars of music, moving toward the climax that is a ghostly rectangle of a whitish-greenish cast containing a pink spectral center, almost an inner eye of the kind Poe anthropomorphized in "The Fall of the House of Usher." These rectangularized shapes disembody the "meanings" known to Rothko in his mythic phase, but they are meanings nonetheless. That is, he has found equivalences to specific feelings of himself in the world and transmits them through stating the paradox: now you see it, now you don't. *Where* is the picture? In the thinnest of membranes covering the threads of the canvas? Hovering before it? Receding behind? In the aura that he sometimes almost literally describes, as when he binds certain works of 1949 with whitish outlines? By 1950, a crucial year for Rothko, these questions were absorbed by an enormous will to work, canvas by canvas, toward transmitting the unnamed passions with which he had lived for so long and for so long sought to express. He had believed, as had the French thinkers, that there was "inter-subjectivity." He could assume that, as he felt the difference between one canvas and the next, so would the onlooker. If he had reduced his composition very nearly to just a few divisions of color, it was color that would be the carrier of mood, of even specific mood. For him, as he always said, each canvas was completely different from the last, despite compositional similarities. For those who, like the phenomenologists, could live in the painting itself, it was not difficult to understand Rothko's elation, or his continuing struggle to make more and more precise the nature of his experience. There were after all other artists attempting a similar enterprise. There was Giacommeti, perhaps the closest parallel, whose quest to locate himself in a space described as "human" brought him to recommence each time with the same configuration, always marveling at how different the nuances of perception can be, and how elusive finally, final definition will be.

Sympathetic viewers such as Thomas B. Hess and William Seitz could find affinities in these new abstractions with certain painters of the past—painters who had attempted, through recapitulating their motifs—to go beyond the given appearances of things. Hess sensitively discerned a Whistlerian sense of space, and Seitz noted that

> Historically, whether or not any influence is involved, Rothko's means might be said to be a continuation of the combination of parallel organization with color films of equivocal depth in Whistler and Monet, though it is a partial or superficial correspondence, for his aims are quite different.[51]

His aims, Seitz felt, always went toward the absolute: "Rothko values the quality of 'immanence,' of a spirit which is indwelling, unified, complete." And, in an indirect quote from Rothko, he designates the Apollonian-Dionysian dualism as the essence of tragic art. "Antitheses, Rothko feels, are neither synthesized nor neutralized in his work but held in a confronted unity which is a momentary stasis." Despite the lush colors of the new works, Rothko had not abandoned his old ideal of "the single tragic idea."

8

"I am still looking for the fabulous," Rothko wrote to the sculptor, Richard Lippold, during his first trip to Europe in the spring of 1950.[52] He had decided to travel to England, France, and Italy with Mell, whose mother had died and left her a small inheritance. Their general economic situation had been poor for several years. Rothko's sales were meager despite his growing reputation. His part-time job at the Brooklyn school could not support them, and he was looking for a job near New York. In his letter he inquires of Lippold about a job in Trenton, New Jersey, and then speaks of his quest for the fabulous, "which they say I will find in Italy." The letter dated May 6, from the French Riviera, hints at Rothko's conflicts. Europe had long represented the spiritual source for Rothko and his friends. But it was also a potential seductress of which the enterprising traveler to the new must be wary:

> I feel like staying put here somewhere for a month or two and making again these things which I am sure few here could have a feeling for: I never realized how really new our world is until I came here.

There were other experiences, though, in Italy, that would subtly invade Rothko's heart and draw him back to Italy twice more in his life. The "really new" had its subsoil of the old, and Rothko was increasingly drawn to it. The fabulous would be found, and not in landscapes but in works of art, much as Rothko often provocatively denied his interest in the work of other artists, particularly other painters, no matter in what century.

Whatever he brought back from that first encounter with Europe (perhaps it was the memory of its architecture; perhaps a first exposure to Fra Angelico, the Thomist enamored of "radiant light everywhere" as Giulio Carlo Argan, an Italian acquaintance of Rothko, would put it), Rothko returned with a reinforced will to surpass. From now on he would court extravagance unabashedly. It is at this point, when he is in his late forties, that he cedes to his will to be ravished. His "enterprise" would take him beyond painting—although painting was his only means—to a point where the *effect* of his painting, just as Mallarmé's effect of words, would be ravishing. What Rothko yearned to possess was some vision of ineluctable beauty that would "tear away," as the word ravish suggests, the uncertainties with which he struggled to the end. The great sensuousness that resided in him and had been so uneasy when attached to things suddenly surged up, when freed of things. His desire to immerse himself (the possibility had been suggested by Nietzsche) in a universe of values rather than things, had found its means. Now Rothko would draw closer to his ideal—the poignancy of music and poetry. The "spirit of music" would inspirit his canvases. He did not forget his Nietzsche. "For if music is really a language of emotion," wrote Susanne Langer, a contemporary of Rothko's, confirming Nietzsche "it expresses primarily the composer's *knowledge of human feelings,* not how or when that knowledge was acquired; as his conversation presumably expresses his knowledge of more tangible things, and usually not his first experience of them."[53] But Rothko was a painter. He had to start again each time to find his first experience of things, without language as mediator.

The boldness of Rothko's address from 1950 on was accompanied by, or perhaps incited by, a stiffening of ethic. If a picture lives by companionship, then companions would have to be carefully chosen. His companion in rebellion, Clyfford Still, was again on the scene in New York when Rothko returned from Europe and the two resumed their old camaraderie. Rothko began to talk of "controlling the situation" and sought to protect his enterprise from the encroachment of a gradually expanding art world, filled with disquieting aspects. Although he would exhibit, once more, in the Whitney Annual of 1950, he was gathering his courage to renounce public exposure in uncontrolled situations such as the untidy mass exhibitions at the Whitney, which asked that Rothko present work to the museum for possible purchase. Declaring that since he had a deep sense of responsibility for the life that his pictures would lead out in the world, Rothko said he would "with gratitude accept any form of their exposition where their life and meaning can be maintained, and avoid all occasion where I feel that this cannot be done . . . at least in my life, I must maintain a congruity between my actions and convictions, if I am to continue to function and do my work." Five years later he reiterated this position to the Whitney's director, Lloyd Goodrich, adding that he understood that he might sound pompous, but asking that his response be seen in the best of faith.

When curator Dorothy Miller chose Rothko together with Baziotes, Pollock, Ferber, Tomlin, and Still, among others, for *Fifteen Americans* in 1952 at the Museum of Modern Art, she found Rothko at first charming and acquiescent. Later he became demanding. He wished to control the installation of his work, even in relation to the others, and he and Still adamantly refused to let their work travel, forcing Miller to cancel plans to circulate the show in Europe. Rothko's unwillingness to appear in group exhibitions, as he told Betty Parsons, was ostensibly because "they just drag you down, while you don't drag them up."[54] Beneath his intransigence, and a need to exercise his will, was an instinct to preserve what he valued most in his work: its potential of enveloping, of drawing the spectator into its ambiance the way Mozart could draw his listeners into his opera in the resounding overtures. Rothko's cultural preoccupations were no less serious and important to him during this period than they had been in his earlier phases. His convictions were essential to his pursuing his work, and he still debated with himself and a few of his colleagues the consequences of his own gestures. During the early 1950s he saw Guston and Motherwell frequently. At times, when he had brooded long enough in his studio, he would set out looking for Guston, who could often be found talking in a few favored bars in the late evenings.

The crucial questions to which Guston and Motherwell addressed themselves in those years of self-definition were questions familiar to Rothko, having to do with the passing-beyond aspect of painting. Both Guston and Motherwell had participated, with a high sense of adventure, in what was felt to be a dismantling of a materialistic tradition going back to the Renaissance. Both had seen the "risk" inherent in painting which turns its back on the named world. Guston was reading intensively: Dostoyevsky, Kafka, Camus above all. Motherwell was preoccupied with Kierkegaard. The conversations these artists had were not about the art world, the world of commodities they so much despised, but about the fate of art itself and the predicament of the artist. The Existentialist bias was evident in a curious obsession on the part of all three of these artists with Kierkegaard's treatment of the parable of Abraham and Isaac. Rothko was accessible to Kierkegaard's argument in his very bones, given his early training and his constant awareness of the great drama in the Old Testament. From Kierkegaard, and from Camus and Sartre with their vision of "authenticity," comes the preoccupation of certain New York School artists with the ethical dimension of painting. Or rather, painting as an ethical act of choice, committed to locating and expressing the deepest of human values. When these artists spoke of "risk" they merely meant to signal their awe inspired by the new conditions with which they confronted themselves in the act of painting. Motherwell would write in his statement for the *New Decade* show at the Whitney in 1955 that pictures are vehicles of passions and not pretty luxuries. "The act of painting is a deep human necessity, not the production of a hand-made commodity . . . True painting is a lot more than 'picture making.' A man is neither a decoration nor an anecdote."

The passionate adherence to a philosophical humanism that characterized Rothko and his sympathetic colleagues made it difficult for Rothko to accommodate less romantic visions of the mission of painting. When in 1951 he landed a job at Brooklyn College, where the department harbored a few non-objective artists who took a practical stance in relation to the training of painters in the basic forms of art, Rothko ran into immediate conflict. His deep antipathy for what he summed up as a "Bauhaus" approach emerged. For him there never was and never could be something called "design" that could be taught. All the exercises with pure forms carried out in the Bauhaus tradition (although *not* the tradition maintained there by Klee and Kandinsky) seemed to Rothko futile. His clash with his Brooklyn College colleagues stemmed from his absolute conviction that painting was not apprehended at all if it were apprehended in terms of design. He told Seitz, "One does not paint for design students or historians but for human beings, and the reaction in human terms is the only thing really satisfactory to the artist." When his attitudes at Brooklyn College became too explicit his fellow teachers turned against him, and in 1954 they denied him tenure. At a hearing Rothko defended himself against their accusation that he was not "flexible" enough, and in his notes reconstructing the meeting,[55] set up in question and answer form, he vented his sarcasm. When asked why he was willing to work in a department devoted to an antagonistic philosophy, he answered:

> That is the history of every artist's life. If we awaited for sympathetic environments, our visions which are new would never have to be invented and our convictions never spoken.

When the questioners state that a department is like a team, he answers

> What kind of a team? My idea of a school is Plato's academy, where a man learns by conversing with men of consequence.

This Rothko believed fervently. Years later when he visited Harvard's Carpenter Center, where a Bauhaus approach to design prevailed, he walked silently through a studio full of geometric paper models. Later, when an older European with a well-lived, sympathetic face asked him what he thought, he leaned forward at the luncheon table, with one of his warmest smiles, and answered: "If I were teaching here, I would have them all do your portrait instead." The company laughed. But Rothko was utterly serious.

By the time the upheaval at Brooklyn College took place, Rothko was well set on his course, although he was still in a constricting economic situation. His daughter Kate, born in 1950, would soon have to start school. The loss of a job, even at a time when Rothko's name had already become important in the context of the rise of the New York School, was a serious event, and Rothko was now over fifty, making other teaching jobs less likely. As irritating as he found the situation, Rothko gained something in anger. He worked all the more in his studio and hardened himself increasingly against what he suspected to be a philistinism even in the way the world now seemed to give acceptance to his work. He opposed himself to his time, and even, to some extent, to the movement of which he was said to be a part.

In 1950, Rothko along with seventeen other vanguard artists had signed a letter protesting Metropolitan Museum policies toward contemporary artists, and had posed for a picture for *Life* magazine by Nina Leen—a picture that became famous and that labeled the artists "The Irrascibles." Essentially these artists were the nucleus of the New York School. By the time Elaine de Kooning published her article on Rothko and Kline in *Art News Annual* of 1958, in which she used one of the terms to identify the Abstract Expressionists—Harold Rosenberg's "action painters"—Rothko was ready to separate himself even from his old companions, writing in a letter to the editor that "real identity is incompatible with schools and categories, except by mutilation." Action painting, he said, was antithetical to the very look and spirit of his work.

Rothko felt a kinship with other artists, but was always fiercely concerned with "real identity." Discussions with Motherwell, Guston, Ferber, Stamos, Gottlieb, and many others during the early to mid-1950s seemed always to come back to the problem of preserving the personal integrity of an artist's statement from the wayward interpretations imposed on it by the combined forces that had come to be known as an art world. Rothko was consistently suspicious of institutions, critics, dealers, and the newly established cast of collectors that, by 1955, would be noticed even by *Fortune* magazine in a two-part article (December 1955). It called its readers' attention to the alluring new field for investment—living American artists—and advised its readers to invest in "pioneers," naming Pollock, de Kooning, and Rothko. Rothko felt that the "subject" of his painting was poorly understood, although he had created his own language, a language designed "to avoid doing violence to objects" as he said repeatedly in the 1950s. Watching the expansion of the art world, he felt a profound distaste. "It has degenerated into a free-for-all," he said in 1956, and added bitterly that in America, "one can never become a patriarch, one simply becomes an old man."[56]

If he longed to become a patriarch, the means were at hand, but it was the wrong culture. America's indefatigable quest for the new and the young would thwart him. Rothko sensed, with some accuracy, that his "language" would reach only a few, while his name would reach many. Increasingly he was exasperated with the written responses to his shows. When Katherine Kuh organized an important exhibition at the Art Institute of Chicago in October 1954, he wrote that he abhorred "forewords and explanatory data" because the result is "paralysis of the mind and imagination."[57] When his work was seen not only in America but in an important exhibition *Modern Art in the U.S.A.* that traveled throughout Europe in 1955-56, and in *The New American Painting* that also traveled from 1958 to 1959, he nurtured his skepticism, not allowing himself to believe in his growing fame. Yet, he remained convinced that he could "communicate something about the world" in his new language.

He tried in 1958, for the last time in public, to specify what it was in the human condition, he wished to express. He spoke of the "weight of feelings" and used a musical analogy: Beethoven and Mozart have different weights. He had told Elaine de Kooning shortly before, "I exclude no emotion from being actual and therefore pertinent. I take the liberty to play on any string of my existence."[58] In doing so, Rothko performed difficult feats of thought. Whatever was proposed to him by way of an explanation of what he was doing, and whatever he proposed to himself, was subjected to endless scrutiny. None of the conventional approaches to painting could satisfy his vision. He disclaimed an interest in color, although color, he conceded, had remained his only means. "Since there is no line, what is there left to paint with?" he told Elaine de Kooning, adding that color "was merely an instrument." When he was called a colorist, he angrily disclaimed it, pointing out that colorists are interested in arrangements, while as soon as he saw his own painting as an arrangement, "it has to be scrapped." When asked about space he was similarly recalcitrant. Space, he insisted to Seitz, like color or flatness, is not essential to his conception. "He emphasizes the material facts: that his variously shaped color areas are simple 'things' placed on a surface." Years later, in 1961, he told a reporter: "In our inheritance we have space, a box in which things are going on. In my work there is no box: I do not work with space. There is a form without the box, and possibly a more convincing kind of form."[59]

In all his statements Rothko was trying to find a verbal context for what he ultimately felt to be inexpressible in words—his consuming love for the virtual which urged him again and again to make a sign that would at once be, and not be. He would repeat the Existentialist notion that, as he put it, "a painting is not about experience, it is an experience." The nature of the experience, of course, had to be in space. What else could a painter qualify? But these virtual spaces he dreamed had different qualifications. Basically Rothko had to assume that there was a space in which others and he could commune. Otherwise, what could a painting mean? Yet it had to be a material fact, "things" on surfaces. If he could submerge himself, and the others, in the spaces he created, the spaces would no longer be virtual. The painting would no longer "seem" but would "be." And yet, it would have the spectral presence that is literally neither here nor there.

Rothko had to assign values, and often his viewers were skeptical. In 1956 when he still worked in a cramped room and could never have more than one or two of his large canvases visible, he used to keep one of his earliest huge abstractions, "Number 22," 1949, against a wall in a narrow storage area. This early essay into a kind of limitless space, with huge areas of floating yellow and orange, interrupted only by a red band straddling the canvas from side to side, shocked unaccustomed eyes. Rothko had not quite reached the ambiguity he would shortly perfect and, to call attention to the picture plane and its function as the final determinant of image, he had scored the rectangular red form with scraped lines. This painting, he said, with its large area of yellow and its bright red was perceived by most people as optimistic. But, he emphasized, "it is tragedy instead." He had assigned a value which in time would come to be understood, but not by everyone. The painting, by its scale alone, could be an equivalent to an epic drama. The eye is given a

24. *Number 22*   1949

desert of yellow in which to wander, perhaps with anxiety, until it reached a narrow border of still paler yellow which with its fading edges is not even really a border. The very time it takes to reach a visual resting point in scanning such a canvas is enough to endow it with faintly disturbing qualities that Rothko could see in terms of tragedy. For him, clearly, the "things," the roughly rectangular shapes moving now slightly forward and now back, were, as he said, actors enacting events in which there were unnamable feelings that would find a resolution on the surface of his canvas.

There were the shadowy whispers behind the final surface, speaking of hidden events (for Rothko admitted, with irony, that his paintings *were* façades, as Elaine de Kooning had suggested). The source of light was skillfully concealed, so that the scanning eye could never be quite sure. Here was a yellow surface, but beneath was Rothko's familiar *vibrato* giving the surface an immaterial visage, a feeling of virtuality. Even yellow, with its conventional association with sunlight, would undergo Rothko's transformation of meaning. It, too, partook of a long harbored and enormously refined love of chiaroscuro. It was Rothko's portion to take what was most moving about the love of light in the painterly tradition. He had been moved not only by Fra Angelico, but by Rembrandt. He would smuggle the great tradition into the 20th century. Where there is a plane in his works of the 1950s there is also a shadow. In each case there would be an underpainting meant to be sensed as shadow, and an oscillating surface meant to be sensed as light. To enhance luminosity, Rothko often resorted to the methods of the old masters. He liked to use tempera, which he concocted with fresh eggs, as a base. When he insisted that it was not color that interested him, Rothko was not being arbitrary, for what he meant was to create light, generate light by overpainting, masking, thinning, and thickening, and working for the musical effect, the *vibrato* to which he responded in the most poignant of Mozart's late works. Reaching for these rare effects, Rothko managed to invent juxtapositions of colored surfaces that had no precedent. Robert Motherwell thought that:

> Rothko's mixture resulted in a series of glowing color structures that have no exact parallel in modern art, that in the profoundest sense of Baudelaire's invocation to modern artists, are *new*. So

new that if Rothko had not existed, we would not even know of certain color possibilities in modern art. This is a technical accomplishment of magnitude. But Rothko's real genius was that out of color he had created a language of feeling.[60]

Motherwell accurately described Rothko's spirit as "a certain colored despair that nevertheless glowed with an inner luminosity of color that is a 'poignant' occasion—to employ one of his favored terms." The language of feeling which Rothko developed through the weighing out of measures of color intensities was far from the demotic language, and depended as much on an occult vision of shadows as it did on light.

These masked chiaroscuro effects unavoidably established moods to which Rothko would give their "weight." In the 1953 "Number 61" Rothko takes his large canvas and fills it so that it brims with presences that seem to inhabit different spaces. As in many of his later works, he weights the painting at the top by placing the brushed, scumbled brownish-red over a blue ground. It is read as both density and transparency, but basically, it is read as a darkness against which the scraped and airy blue horizontal beneath it plays, opening out into an azure of infinity and seeping into the darker blue below. These weighted and delicately balanced densities are replete and, in filling the canvas, have a kind of lyrical grandeur. But in another painting keyed to blue, "Whites and Greens in Blue" (1957), the feeling is utterly different. There is little exuberance here. Rather, gravity. Fate. The three forms lying on a blue ground have a kind of finality. The murmuring underpainting is controlled, held within the tightly organized central scheme in which Rothko suggests impingement, but barely. At the time this painting was completed, Rothko claimed that he was creating the most violent painting in America. When I visited him during 1956 and 1957, he always brought the subject up, without offering further elucidation. I took it to mean that by a supreme effort of will he had harnessed turbulence and was painting the paradox of violence; that the colors that produced immeasurable tensions among themselves were conceived as symbols. They had been a thousand times refined, and all smudgy echoes of the everyday world had been removed, but for him, and then for me, they were equivalents of complex emotions.

Lightness and darkness, yes. But not color with any conventional designation of meaning. What could it mean, for instance, when he placed a large horizontal of dark blue over a red-orange ground, as in "Blue Over Orange" of 1956? The burning red is all but blotted out by the depth of the blue, but then, it is modified. Above, the red surround is fierce, below, where it is given back to the sun and is a pale orange expanse, it is almost tender. The brush strokes soften contours; ravel their edges, and at the same time, paradoxically, give a certain relief quality to the form. In the same year Rothko painted "Brown, Black on Maroon," with an entirely different "weight" of feeling. When he moved to the darker register, it was inevitable that the tragic associations so important to him were invoked. The chiaroscuro is heightened. A whole drama seems to be taking place behind the thinly disguised upper register which is pushed into a recessive position by the brilliant shivering band of scarlet light beneath it. The largest form—the black made so rich in its thinly layered overpaintings—then comes forward, an emphatic "thing" with a weight and stability that cannot be altered by the flickering reds of its nimbus.

The suggestion of aureole is not exaggerated. Many of Rothko's paintings recall his earlier obsession with the aura—the subtle, invisible emanation or exhalation; that cloud of air that hung about the gods. In the 1940s he was already endeavoring to paint the suspended, infinitely extensible air that hung about his mythic visions. If he was to find the doorway of which he spoke, leading beyond the everyday, he would need to be able to conjure his aura. The idea of an aura is precisely that it must be more than perceived. Paradox is in its nature. All perception is paradoxical since we can never be sure that we have "seen" all that is there, and in some way we select aspects that are forever changing as we see them, as Cézanne discovered. If our habit is always to carry such a paradox in mind, we will, almost by reflex—a *cultured* reflex—perceive this aura even when we are not in its presence. Rothko was right to count on our perceptual culture. For us, his auras can be like music in that we carry with us the after-image of his generalized auras. He said he wanted to hold things in suspension. He wanted to maintain both the immediate experience (which he called materiality) and

25. *Number 27*   1954

its description, which necessarily moves away from it. For this he invoked another paradox, which is that of time. His painting was to be immediately perceived, and yet, to unfold its communication in time. Light from outside would slowly reveal the light within. A slow rhythm of apprehension would be established. Those whites which Rothko had made into a material thing, having the weight and value of color, were to serve as metaphors for the passing-beyond of the thing, but they were also the thing. In 1951 he painted "Number 18," an immensely bold image in which a great plane of dense white rivets our attention and then, slowly, filters out into the reddish edge, as though it were given, only to be dissipated. Its shadow lies below, mitigated by the hints of modeling in its horizontal bar that suggests how near the ideal blankness the form above has come. This reciprocity is always present in Rothko's paintings of auras and it is wedded to the question of metaphor in painting. It is never light—that ever-vanishing virtuality—as in the sky over the sea, because it is material, it is paint. But it is also a most precise metaphor of such light. At times Rothko went further with a romantic vision of aura and specter, as in the 1954 "Number 27" in which the aura itself has an aura, and all things are doubled and mirrored, with whites beneath the central white. The reversibility or reciprocity here is provided by Rothko's thin technique which allows darker shapes to read as light and the white to read as a denser substance—some ghostly reminder of another place, perhaps the place in which Rothko once said he could breathe and stretch his arms.

These works of the 1950s leading up to the mural cycles were often very large, not, as Rothko said, with the intention of monumentality but with the intention of intimacy. He was still a Nietzschean. He wished for himself, and by extension, for his viewers, an experience of dissolution into a whole not unlike Nietzsche's; also not unlike Mallarmé's who had said toward the end of his life that the sole duty of the poet was to give an Orphic explanation of the universe. The Orphic is dramatic by the very nature of the myth of Orpheus and sometimes, in these large expanses with their grand curtains of darkness, as in "Number 128 Blackish Green on Blue" (1957), Rothko sounds his strongest intimations of origins. There is

26. *Number 9*   1958

a silence here, the kind of silence Melville called "the general consecration of the universe." In "Number 9" (1958), with its play on hues of red—darkened with browns in one register, heightened to the pitch of fire in another—Rothko offers in his tacit language an equivalent to a complex of emotions that can be likened, as he knew, only to the equivalents produced by poets and composers.

The music analogy, or rather, the concern with the "spirit of music," was simply a way for Rothko to deal with the unprecedented nature of his enterprise. He sought out musicians and he was friendly with several composers, among them Morton Feldman with whom he used to visit the Metropolitan Museum. Feldman maintained relations with several painters in the Abstract Expressionist group, and was always a willing and demonstrative viewer. His own work in the 1950s was characterized by his friend and sometimes mentor, John Cage, who offered in The Club in 1949 "A Lecture on Something." With his usual mystification, Cage had called attention to the sense of void typical of Feldman's work of the 1950s. "When nothing is securely possessed one is free to accept any of the somethings," Cage said. "It is quite useless in this situation for anyone to say Feldman's work is good or not good. It is."[61] He added that Feldman's music "continues" rather than changes. In the same sense, Rothko's paintings of the 1950s continued, with each canvas expressing in its tacit language an aspect of his vision of the entire human drama; of the single idea that would represent all the ideas of human feelings.

9

When Rothko made his last public statement, on October 27, 1958, in an old classroom at Pratt Institute, he began speaking without his notes. He spoke haltingly in a quiet voice about what it meant to be an artist, pointing out that he had begun relatively late in life, and that he had formed a speaking vocabulary long before he had formed a painting vocabulary. It was easier to paint pictures than to talk about them, he thought, but still, talk was necessary. He issued a rebuke to Expressionists who "strip themselves of will, of intelligence, of the bonds of civilization" and who seemed not to understand that "painting a picture is not a form of self-expression. Painting, like every other art, is a language by which you communicate something about the world."[62] The dictum "Know thyself" is only valuable he said, if the ego is removed from the process in search for truth. Not surprisingly, he invoked Kierkegaard's *Fear and Trembling*. He had pondered the parable of Abraham and Isaac for years, and frequently in conversations with friends had cited it, for he was seriously engaged, as had been Kierkegaard, in unraveling the meaning of that Old Testament horror story.

Rothko's thoughts on the Kierkegaard interpretation of the Old Testament story went something like this: Kierkegaard is describing the artist. Abraham is called upon to commit "a unique act which society cannot condone." Abraham is caught between the universal law and the law that governs the individual conscience (much as the heroes of Rothko's other admired source, Aeschylus, had been). The artist—that is, Rothko—is compelled to sacrifice, to commit a unique act, as all artists must, even if it comes into conflict with the universal law. Behind this interpretation lay the tacit hope of retrieval, even redemption. It is not too much to talk about "faith" when speaking of Rothko, as skeptical as he sometimes wished to appear, as modern as he professed to be, as detached as he claimed he was. "The fact is," Kierkegaard had written, "the ethical expression for what Abraham did is that he wanted to murder Isaac; the religious, that he wanted to sacrifice him. But precisely in this contradiction is contained the fear that may well rob one of one's sleep." Rothko, as close friends knew, was the victim of robbery almost nightly. The ethical imperatives of his youth gave him no peace. Old friends remember the animated arguments of the 1930s when Rothko would pose the Dostoyevskian dilemma, asking—for instance—if there were a fire, and in the house was a child and a Rembrandt, which would you save? His answer was always a passionate "The child! No painting is worth human life." His dwelling in the moral realm gave him no peace. (Once we enter this realm in which Rothko spent much of his time, there can be nothing but difficulty. Kierkegaard says as much in his *In Vino Veritas*, arguing that the "loveable" cannot be defined. He finds his interlocutor's position untenable since "it contains a double contradiction—first, that it ends with the *inexplicable,* second that it *ends* with the inexplicable; for he who intends to end with the inexplicable, had best begin with the inexplicable and say no more, lest he lay himself open to suspicion. If he begins with the inexplicable, saying no more, then this does not prove his helplessness, for it, anyway, is an explanation in a negative sense; but if he does begin with something else and lands in the inexplicable, then this certainly does prove his helplessness.")

In Kierkegaard's sense, Rothko's was a religious enterprise, with the sacrifice and attendent violence. There was always the helplessness before the inexplicable, and the uneasiness in the interpreted world. Still, Rothko could dream of the symmetries and grand dramas of the ancients, and once again, passed beyond the contingent when he stressed in the 1958 lecture going beyond individual ego. Art is not self-expression, as he had thought in his youth. The notion of schools of art stems from an idea of self-expression that regards the activity of painting as a process in itself, he said. But he could not accept that. "A work of art is another thing." The notion of self-expression, he said, was proper to the vanity of a beautiful woman or a monster, but not an artist. "For an artist, the problem is to talk about and to something outside yourself."

For several years Rothko had been moving toward the views he expressed in the Pratt lecture. His desire to immerse himself in the spaces his paintings proposed became more and more imperious until it occurred to him that the most satisfying means would be the most literal: that canvases would surround the viewer as murals. Fortunately the opportunity to move into the larger scheme—what he called the "jointed scheme"—arrived in the form of an invitation by Philip Johnson to paint murals for the Four Seasons Restaurant in the Seagram Building. Rothko was spurred to find a large studio, a former YMCA gymnasium at 222 Bowery, which he set up in the spring of 1958. The following spring he set out for Europe, where he was well known thanks to the exhibition of his work in the previous year's Venice Biennale. He was treated like a king, as he later reported, and he obviously enjoyed the affectionate attention he received in Italy. By the time he got to Rome, where old friends such as Toti Scialoja and Gabriella Drudi were awaiting him, Rothko was in high spirits, eager to look about him, talk, meet people, and seek experiences with artists of the past. They visited the oldest churches in Rome, and drove out to Tarquinia to see the Etruscan murals resplendent with the colors that Rothko loved best. The scale of the chambers in which the Etruscan painters had so elegantly praised life was the human scale that Rothko often spoke about, having little to do with literal size (the rooms are small) but with the way experience is presented, without hyperbole and heroics. In Rome itself, Rothko wandered about with his friends, taking in the horizontal architecture and discussing its unique charms. His interest in architecture, as Drudi says, was greater than most people thought. When the Scialojas had visited New York two years before, Rothko had taken them almost at once to see the Sullivan building that few ever bother to visit in downtown New York.

146

One of the most significant events during that important trip to Italy was a visit to Paestum and Pompeii. As a frequent visitor to the first floor of the Metropolitan Museum, Rothko had long been moved by the Greco-Roman vision. But the visit to the actual sites was immensely moving to him. He spoke of it many times to friends, and John Fisher, an editor of *Harper's* magazine who had met Rothko on the ship going to Europe, and again on the trip to Pompeii and Paestum, captured some of Rothko's excitement in his account.[63] At Paestum, a remarkable site where the Greeks had built a majestic Doric temple whose columns still bespeak its singular beauty, he was asked by some Italian tourists if he had come to paint the temple. "I have been painting Greek temples all my life without knowing it," Rothko replied. Fisher also mentions that Rothko found "affinities" in the House of Mysteries in Pompeii—affinities he had already experienced at the Metropolitan Museum with the wall paintings of Boscoreale. It was during this second sojourn in Italy that Rothko again visited Florence, where he saw two important monuments with a different eye. The first is Michelangelo's Laurentian Library with its sense of closure; its stairwell surmounted by blind windows, its corridors hermetic, as was suitable for a monastic library. Fisher mentions Rothko's allusion to Michelangelo's walls in the staircase room in the Medicean Library. A few months later Rothko spoke of the corridor of the Laurentian Library which he felt had lingered in his memory as he worked on the Seagram murals. The corridor, with its insistence on enclosing the presence moving through it, was a likely metaphor for Rothko who, even before the murals, talked always of "controlling the situation."

The other important experience in Florence was Fra Angelico's murals in the Convent of San Marco. Rothko's apprehension of Fra Angelico was not casual. When he was moved, he could be very thorough in exploring the causes for his emotions. In Italy, he not only spent careful hours perusing Fra Angelico's works, but he also discussed the painter with one of Italy's most acute art critics and historians, Giulio Carlo Argan, who had recently written a monograph on Fra Angelico. This painter was to remain a beacon for Rothko. He would return to San Marco during his last Italian journey in 1966. It is not a matter of "influence" or even temperamental affinity. Fra Angelico was important to Rothko because Rothko understood the context within which he functioned, and because he himself had shifted his sights. His aesthetic was now a renunciation of self-expression in favor of meditation. His innate Platonism had triumphed over the tumultuous emotion that had once governed his works. Quite possibly Argan's interpretation of Fra Angelico's unique situation in 15th-century Italy helped Rothko resolve his point of view as he embarked on his most important artistic adventures—the mural cycles of the late 1950s and 1960s. Argan traces Fra Angelico's painting career, stressing always that he was a churchman and theologian who rose in the Dominican ranks and who was attuned to intellectual values. He was not the naïve, rapturous, and angelic figure of mythic art history. Rather, he expressed his point of view through "the purely theoretical values he attached to light."[64] In the beginning, Fra Angelico took a middle position betwen the two views prevailing in his time. Cennino Cennini, whose views often reflected medieval studio practises, maintained that painting was "the art of eliciting unseen things hidden in the shadow of natural ones . . . and serving to demonstrate as real the things that are not." Leon Battista Alberti, a modern humanist with a scientific bias, maintained that "invisible things cannot be said to come within the painter's compass and he seeks only to depict what he sees." Fra Angelico's mission, as a Dominican monk, was to depict what he saw in nature in the light of its unseen source, God. He and his order were resolute Thomists. When Fra Angelico painted his scenes of Edenic beauty, he was true to the Thomist vision of Beauty as "that in which the eye delights," but he acknowledged the Thomist principle that painting is knowledge as "it satisfies our desire to understand and know." For the Thomists, the world and its beauties could be depicted only as effects for which there could be only one cause. God would be the source of all visual pleasure, and the light that would grace the world of nature would always flow from Him. Therefore, in Fra Angelico's exemplary panels, the light is evenly distributed, not modified by Alberti's new principles of perspective, which Fra Angelico used intermittently for his own purposes.

The teachings of Blessed Giovanni Dominici lay behind Fra Angelico's mission. Dominici had laid the groundwork for Fra Angelico's order in his approach to nature as tangible proof of God's goodness. He reasoned that if we are to contemplate God, we must scale the ladder of logic from effects to causes; indeed, to the prime and unique cause where at last, "the thirsty intellect is quenched" and the order of the natural world expands to the universal order of the heavenly hierarchy:

> Here are the rejoicing of angels and apostles, the dancing of martyrs and confessors, the choir of virgins, the joys of all elect. Here are the true sun, the morning star, the flower of the lofty field, the lily of the valley of the just, the rose that never fades, the violet that never withers, carnations, cinnamon and balsam, with the most fragrant perfumes of the Kingdom of the Blessed. Every pleasure has its source in God. There is no pleasure that comes not from Him. How foolish is he who seeks pleasure elsewhere![65]

Many of Fra Angelico's resplendent middle-period works are seemingly illustrations of Dominici's lyrical instructions. Light would be the binding vehicle of meaning, making all the objects and figures instinct with its source. As Argan writes, "Fra Angelico posited light on the qualitative principle by which human experience, limited in scope and heavy with 'quantity' might be sublimated into a supreme ideal of being."

But these altarpieces and predellas in which Fra Angelico lavished heavy pigments, literally jewel-like, as when he uses large amounts of lapis lazuli, and in which there are many allusions to nature, are different from the frescoes that so moved Rothko in the Convent of San Marco. Here, Fra Angelico divested himself of the world with its cheerful colored substances. He sacrificed the lapidary beauty of heavy pigment. Instead, he moved to sober transparencies in which there are only traces of the past splendor in which he had decked the world. These murals in the cells of monks and in the cell of their patron, Cosimo de Medici, are not meant to edify or cajole. There can be only one reason for their existence: to assist the monk in his ritual of contemplation and prayer—activities that can lead to rapture, but only within the monastic walls. Argan remarks on "the thoroughness with which, at San Marco, history, nature and myth have been ruled out in favor of rite and symbol." Fra Angelico limits his themes, using the symbols of Christ's Passion sparingly. He scales down his colors to their faintest values, taking advantage of the soft white ground of fresco to lend an evenly flowing, unnaturally pale light that emanates from no solid source and illuminates no solid body. Everything in these works is reduced to its purest essence, and often there are whitenesses that bespeak a totally unearthly experience. There is no given space here, no space that can be likened to perspectival perception. The single figures are consumed by the cool flowing light as are incidental details—a briefly sketched hill or tree, or a blue ground that cannot even be perceived as sky, so much has it become a symbol for something other. Anyone moving from cell to cell in San Marco would be struck by the evanescence, the immateriality of the scenes depicted, and by the way Fra Angelico conjures a space that knows no limits and yet flows tangibly before one's eyes. The surprising note of tragedy in the sober transparencies and the starkness of presentation are also sustained, as though Fra Angelico, among his peers, allowed himself to speak of another kind of human experience—the despair of the individual seeking his moment of enlightenment; the moment to slake his intellectual and spiritual thirst. Rothko, moving from cell to cell, and feeling the immense emotional investment of this monk who was able to sustain the sanctified atmosphere throughout the large monastery with a stark simplicity of means, felt an emotional affinity that would express itself shortly after in his own mural cycles. He certainly retained the impression left by the large "Crucifixion" in the main hall. Here the three crosses are stark and mournful against a rust, or dried-blood red sky (possibly originally meant only as a ground for lapis-lazuli), and the semicircular composition falls into three bands of light. Practised as he was at seeing the grand scheme over the narrative, Rothko was moved by the austere power of this work.

At the time he was working arduously on the Seagram commission, he was having an intense debate with himself about the meaning of art. He sought out friends who themselves were given to searching questions. There were long philosophical evenings. One of his most stimulating associations was with the poet Stanley Kunitz, who saw Rothko at the time as "a primitive, a shaman who finds the magic formula and leads people to it."[66] Kunitz wrote about poetry during that period that "it is not concerned with communication; it has its roots in magic, incantation, and spell-casting." Kunitz shared Rothko's vision of the contemporary world as fraught with distressing problems, believing, as he said in a public lecture in 1959, "we do not have a steady gaze on a fixed reality."[67] The great work of man, Kunitz felt, is "to set up a world of values to shore himself against the ruinous dissolution of the natural world." He and Rothko had often discussed the moral dimension in poetry and painting. Kunitz saw a danger in the generalizing tendency of abstract painting, quoting Blake's admonition to pay attention to "minute particulars" and Hölderlin's belief that nobility in art lies not in the noble but in the commonplace. Rothko, who had always maintained that he was no mystic but a materialist, warmed to the argument, defending his move away from objects. Kunitz's position was never far from Rothko's, for he posited the ethical as the very core of poetry. "In the best painting," he said, "as in authentic poetry, one is aware of moral pressures exerted; an effort to seek unity in the variety of experience. Choices are important. Moral pressure exists to make right and wrong choices." Even more, he felt, as did Rothko, that "all failures are pardonable except the failure of conscience." They agreed that art constitutes a moral universe. As Kunitz put it: "The world is defended by art, made more livable because art does deal with pleasure. It gives us the image of Eden, the idea, the beautiful. Art, by setting up the ideal makes man discontented with less than the ideal." In an even more specific statement, suggesting deep affinities with Rothko's views, Kunitz wrote in a letter:

> My predilection, however, is clear enough—it's for Insight that subsumes Sight, for Vision that annihilates Taste. The only worlds that count (or do I mean Words?) are self-contained universes. Adam became the father of poetry when he gave names to the beasts of the field. But it was Noah, when he admitted them two by two to the ark, who became the first critic.[68]

Such convictions stated from many sides in the wandering conversations between painter and poet fed into Rothko's enterprise. They gave him the breathing space, the passing-beyond the petty trivialities of day-to-day art world preoccupations, and the confirma-

27. Untitled   c.1944-46

tion of his intuitions. Moreover, Kunitz was attuned to the larger philosophical issues of the postwar period, the issues so avidly discussed in Europe but less often examined in the United States. Years later, Kunitz would look back on his experience with Rothko's work and say that he was moved by its grandeur, its rhetoric of color. "I felt definite affinities between his work and a kind of secrecy that lurks in every poem—an emanation that comes only from language." This restatement of the old dream—the Mallarméan ideal—if uttered, as it surely had been, during the discussions between Rothko and Kunitz, could console Rothko in his task; a task that he labored with incessantly. It is not incidental that Rothko produced several cycles for the Seagram commission, seeking always to capture the "emanation" that for him could come only from paint.

Rothko prepared himself for his task by returning to a medium he had found responsive so many years before. He made a number of gouache sketches in which he allowed himself sensuous play with color, and in which he experimented with sequences of shapes that implied a third dimension more emphatically than his recent large paintings. Some of these sketches initiated a movement on lateral lines, climaxed by images of fiery doorways. Some had bright blue gateways to a relatively atmospheric ground. In some there were yellows and crimsons, and flares of flame blue. It was evident that he was bent on moving beyond the stately containments of his last large paintings, imagining perhaps a "corridor" of corridors through which his viewers would enter Rothko's universe. These initial pictorial ideas with their occasional swelling forms; their clear preoccupation with rousing the senses, and their departure from the symmetries of his earlier abstractions would go through many transformations as Rothko ordered and re-ordered his processional forms. The sequence of his procedures would be, as he had said years before, toward clarity—a clarity he felt he had achieved only after dozens of essays.

In the first series, Rothko stayed close to his sketches, using color contrasts and forms freed from rectilinearity. He stressed vertical elements, no doubt because of the architectural nature of this enterprise. Perhaps the association so many have had with portals on seeing these forms is well-founded. Originally, the dining room graced with an entire long wall of his murals, was to have been seen from an adjoining employees' dining room through large doors. He worked painting by painting, holding in his imagination the effect he wished to receive when they would finally find their form as an ensemble. He had often spoken of his life's work as an ellipse, or sometimes as an oval. It was characteristic of him to think of beginnings and endings as consecrated moments, and his insistence in 1958 in his Pratt lecture that the artist must have a "clear preoccupation with death" and that "all art deals with intimations of mortality" and that "tragic art, romantic art deals with the fact of man born to die" would play its part in these and subsequent mural cycles. Rothko painted, canvas by canvas, a cycle in which the parts would take their place in an imaginary space that he carried about in his mind's eye, and only later in the place they were meant to embellish. (The fact is that none of the "places" allocated for his mural cycles was ever quite satisfactory, and perhaps none could ever be, given Rothko's peculiar fusion of architectural tact and painterly individualism.)

As Rothko worked his way into the second cycle, bringing back with him his memory of experiences in Italy, he strengthened his allusions to post and lintel. The reds and blacks of Pompeii seemed to haunt him. Yet, he meant to talk about human respiration: an organic softness sometimes invaded the rectangular shapes. They hover, trembling, and sometimes seem on the verge of collapse. The original essays in gouache are remembered in the brushwork that in the murals often takes on an expressive function, enlivening or quieting surfaces. His animistic approach at times led to the formation of two entities huddling together. Sometimes there are darkling canvases in which the artist clearly sought to suggest black mysteries behind the final plane of his canvas. The implication of a third dimension is always there, and there were, in several of the paintings of the second cycle, interior auras: there would be the gate-like vertical form in the central recess and within that, a tympanum of shadow.

In the spring of 1958, I had made one of my periodic visits to the studio and Rothko talked again about "controlling the situation." By "situation" he meant many things. He meant to reinvoke his ethical position announced in the early 1940s. He meant, literally, situation—*where* things are situated and *how* they are to be perceived. With the mural commission, his "jointed scheme" would control the situation, although paradoxically, as he had once pointed out, the large canvases take you into them, you are not outside of them, "it isn't something you command."[69] By 1959, he was deeply immersed in the problem of making his scheme conform to his inner vision, working indefatigably. In October 1959, I again visited the cavernous studio with its deep-dyed red floor. Rothko had no lights on, and the great space was dim as a cathedral. When a tiny white cat sprang into the center of the room it was a shock. I felt as though I had walked into a theater, or into an ancient library. The only perceptible object in the huge space was a table, very small in its isolation. Rothko watched my reaction as I examined the arrangement of large canvases and said, "I have made a place." For me it was a place like Tintoretto's place in Venice when I had first seen it, with all the large paintings sunk in darkness so that I would never forget their presences. The paintings I saw that day included a deep maroon with a collapsing crimson rectangle that seemed to sway like a censer in its ambiguous space. At the end of a sequence that Rothko had arranged that day for my visit, a painting with an orange-red slit in an oxblood matrix reminded me of a shape of a Renaissance fortification, or shapes in the Japanese Noh theater. It was a long visit, with intermittent conversation, and at the end, as I was taking my leave, Rothko said: "They are not pictures."

In an important sense Rothko was right. They are not pictures. They are not icons either. They are images, however, if by image we mean that which is thrown upon the imagination and cannot be expunged. Painting was Rothko's only means, but the image was his vision, and who, even the painter himself, can define a man's vision? In the murals Rothko's frankly theatrical use of fiery reds, with their climax of pure pigment, moving with the flickering unevenness of candlelight invokes the blood and fire purification of old ritual. The burning quality was heightened deliberately as Rothko mixed raw pigments into the final surfaces of these canvases. The undeniably portal-like or window-like shapes are not the stony portals of real architecture, but rather the vanishing, never-to-be entered portals of the Old Testament—the Law, the Law as Kafka described in his memorable allusions to portals. The study of proportions in Rothko's mural cycles, which was one of his deepest interests, was not so much rooted in the practical needs of the Seagram restaurant as in Rothko's need to understand the abstract notion of proportion as he might have encountered it in Aeschylus.

These paintings had to respond to an inner vision that could never be quite satisfied in actual *situ*, as confirmed by Rothko's eventual refusal to deliver them. Various reasons have been offered, including his indignation that the original plan to have employees see the works was thwarted when it was decided to change the architecture. More likely Rothko had installed his works in an interior theater of his own with which no real place would ever compare. The history of that last cycle bears that out. When Sir Norman Reid, Director of the Tate Gallery, visited Rothko in 1965 with the proposal that a special room be allocated in the Tate for his work, Rothko saw another opportunity to "control the situation." He rejected Reid's proposal that he give a "representative" group of paintings, offering instead a group of the Seagram murals. He had great difficulty, however, making up his mind, and the negotiations went on for years. Rothko went to London in 1966 to see the space, and, as Reid reported, "like Giacometti," who had also been offered a room, "he admired the scale of the rooms and the light, which can be particularly beautiful."[70] When he returned to New York he wrote to Reid:

> It seems to me that the heart of the matter, at least for the present is how to give this space you propose the greatest eloquence and poignancy of which my pictures are capable. It would help a great deal if there exists a plan which you could send me. I could then think in terms of a real situation and it would make something concrete, at least for me.[71]

After his serious illness in 1968, Rothko wrote Reid that he had had to neglect many things, but that "I still hold the room at the Tate as part of my dreams." He was concerned with every detail and after all was arranged, sent Reid a swatch of the color of his own studio walls so that everything could be exactly as Rothko had worked it out for himself. He had also indicated to Reid that he wished to hang his work in the older section of the museum, near the Turners. Eventually, the nine paintings were hung after his death in a room adjacent to Giacometti's room.

When the Tate paintings are viewed as an ensemble, it is possible to see that Rothko's "jointed scheme" with its ambitious intention of enveloping artist and viewer in a suite of related moods is not entirely without precedent, and that Rothko has moved within a tradition that has already conditioned both the artist and his viewer. The necessity of reciprocity was essential in Rothko's way of thinking about his mural enterprise. If he moved one canvas into a relationship, and then substituted another, it was not that he had no real mural intention. Rather, he sought the enveloping sensation that can be generated when tones from one surface echo another; when the events on the canvas have the psychological effect of drawing the viewer into its space, rather than allowing him to stand outside, as did the Renaissance artists. This almost metaphysical notion of reciprocity, celebrated in the works of the Symbolists at the end of the 19th century, would find its earliest complete statement in the astounding late works of Monet. Clearly, Monet wished to immerse himself in his universe of light and water, with its fluid exits and entrances in our intermittent vision. There was always for Monet the graze of light that would emerge and disappear from one canvas to another. His "Nymphéas," so perfectly ensconced in the Orangerie, could easily be perceived in a different sequence and perhaps elsewhere, for their correspondences are interior. This occurs even in his more specific paintings, such as the rows of cypresses that fill up a space, then another and another space, but always a space that has ambiguous lineaments. Rothko may have thought of Monet. In Monet the idea of transcendence, with its complex paradoxes, is explicit: Monet sits in Giverny, waiting to entrap his moment. Monet works in a scale in which he must invoke time, and live time as he travels from side to side on his canvas. Monet gives himself to the water and tries, like an alchemist, to change the heaviness, the quantitativeness of pigment and its oil binder to the fluidity of air and water. Monet transcends his vision and creates a little universe of elements. Each day he begins again. Each day his painting ritual is to approach the same motif and to wring from it equivalents to his emotional furor. He too wished to dissolve himself, or his self, into a whole, a universe whose silence reverberates.

Despite the fiasco of the Seagram affair, Rothko was eager to find again an opportunity to create his own environment via painting in a public place; a possibility of "translating pictorial concepts into murals which would serve as an image for a public place." Since Rothko always chose his words carefully for publication, his diction here is important. He starts with pictorial concepts—individual paintings. They are to be translated into murals (his love for the prefix -*trans*). These murals are plural but they will serve as *an* image for a public place. In other words, the spectator will come away with an after-image that is whole, a harmonious whole that creates a unique atmosphere in the imagination, as would a symphony on the morrow of a memorable concert. Soon after Rothko completed the Seagram works, Wassily Leontief, a Nobel Prize winning economist who had admired Rothko and been a friend for some years, approached his colleagues of the Society of Fellows at Harvard University with a proposal that Rothko be commissioned to create murals for their future quarters. Most of the Fellows, as Leontief remembers with gentle amusement, were baffled by Rothko's work, but with a little persuasion succumbed to their chairman's enthusiasm. The murals were intended for the penthouse of the Holyoke Center designed by José Luis Sert. Rothko went through the ceremonious interview with Harvard's Fellows with considerable aplomb, and even his interview with Harvard's president did not unnerve him, much to Leontief's relief. Here too, however, there was to be a snag. Before Rothko's murals were installed, the Harvard Corporation decided to use the penthouse not for the Fellows, but for a dining room for official entertainment. Rothko's calm response to the preliminary maneuvers for his Harvard commission reflected his general attitude in the late 1950s and early 1960s. He was feeling far more confident and had experienced several moments of inner satisfaction, both in his works and in their reception. In 1959, for instance, he had been at the cocktail party offered by Monroe Wheeler in the Museum of Modern Art's guest house in honor of Miró. Rothko had approached Miró to say how much he owed to him, and Miró had turned, smiling, and replied, "And I owe a great deal to *you*, Mr. Rothko." Another event—the installation of his own room in the Phillips Collection in Washington—also brought Rothko satisfaction. Probably even the moment when he refused the Guggenheim International Award, writing James Johnson Sweeney that he looked forward to the time when "honors can be bestowed, simply, for the meaning of a man's life's work," also bolstered his self-esteem. Rothko began work on the Harvard cycle in 1961 and finished in 1962, bringing some of the pictorial concepts broached in the Seagram murals to fruition. He spoke, at the time of their installation, of the "oval" character of his enterprise (calling to mind the romantic Herder's allusion to the ellipse: our planet, Herder wrote, moves always around the two foci of our ellipse, knowing and feeling, and is seldom equally near each). In the Harvard cycle of five panels, Rothko's planet moves magnetically toward the pole of pure feeling, but there is a steady reference to knowing, or to a concept. His intention was to suffuse a given space with an atmosphere created entirely by the light and attendant shadows residing within his paintings. He could be in complete command of the emotional responses of his viewers because not only would he control the position of his paintings in relation to the space within which the viewers circulate, but he also controls the definition of the room. From any point in the room, which at Harvard was far from congenial to Rothko's purposes, the paintings surround and work on the sensibilities of its inhabitants. The room Rothko had to accommodate was awkward. Two of its walls were glazed so that the room itself became little more than a corridor between large horizontal windows. Details, such as heavy glass doors and clumsy louvers hiding air conditioning units were intrusive. Yet, they were canceled by the paintings, which managed to expand and engulf despite windows, doors, and heavy dining chairs that partially obstructed the canvases.

Rothko did his best to calculate the light in his "corridor" of paintings, knowing that time—one of the important ingredients in his work—would collaborate with him. These representations of sequences of feelings that have what Coleridge called "companionable form," as does the flame in the hearth, reveal themselves in time. The major triptych on the long wall tended to change character hour by hour. In the midday light the central horizontal canvas with its plum reds reading through blacks and its irregularities of form and value took on a silvery effulgence. The same three can-

28. *Triptych*, Panels 1, 2, 3 of the Harvard Murals   1962

vases in the late afternoon became a densely textured, mysterious mass of cryptic shadows. To the right, the wine red forms of central verticals seem to curve inward, with depth given in the blue-gray tones that can be read below the surface. In the oblique light of waning day this panel with its central rectangular forms feathered on both inner and outer edges pulsates in long, slow rhythms. Contrasting with the langorous mystery of the darker panels is the leftward panel. Because of our reading habit, it is the first of the sequence, but could as well be the last. A fiery orange-red form is suspended like a flaming hoop in purple space, an apparition that appears in its own theater with its own transforming inner stage lights. It too is susceptible to time, and as it is contemplated, the low rectangular knot—the plaquette suspended on a horizontal line that is repeated top and bottom in each painting—becomes a glowing coal, with thicker brush marks and vermilion splashes like lambent sparks. In these paintings Rothko is concerned with conventional values of chiaroscuro posed in unconventional terms. Like many painters in history, he is deeply concerned with the almost indefinable areas between obscurity and light—a concern that would later become absolute. Depth is assured in various ways. There is never a single opaque surface, but a membranous, respiring foreplane image through which several levels of light and shadow announce themselves. Undercurrents are the very heart of Rothko's image. His depth expresses itself also in the way there is never a straight edge—the obvious means to assert frontality and hold to the plane. He takes care to make his roughly rectangular forms ambiguous, always tapering the edges and sometimes grading them off into four or five values so that, symbolically, the viewer can read far back. He used his brush to ensure such reading, as in a red and white, squarish panel in which the white has pink-to-gray intonations that constantly change. The brush established stages in a space that is, as always in Rothko, not exactly circumscribed, described, or bounded. Each painting in this mural cycle has a heart, and each has a pulse dictated by that hidden heart.

By the time Rothko had completed the Harvard murals, he had arrived at a position of considerable importance and was generally regarded as a master, even in the public press. Yet the meaning of his life's work had been rarely discussed in the art press. Rothko had no single powerful advocate, although several extremely sensitive appraisals had appeared during the 1950s. The earliest article dealing with the abstractions was by a West Coast painter, Hubert Crehan, who had acquired his interest during his days around the California School of Fine Arts. Crehan understood long before the others that Rothko was engaged in an undertaking that could find little expression in words. He wrote in 1954 of Rothko's "wall of light" and pointed out that "This rational attempt to destroy the line is undoubtedly based on Rothko's intuitions of the essential oneness of things—a kind of visual metaphor of the unity and integrity of life, consciousness and the universe."[72] In his enthusiasm, Crehan referred to Rothko as a prophet come down from the mountain—which was enough to set Rothko against his interpretation, jealous, as he always was, of his reputation as a "materialist." The next cogent interpretation of his work appeared only in 1957 when Elaine de Kooning wrote "It is no accident that a painting by Rothko is a façade, almost as though his art were trying to hide behind itself."[73] This response also met with Rothko's disfavor, despite his having "sat" for the article at de Kooning's studio, as she wryly reported. Rothko objected largely because de Kooning had lumped him together with others in the New York School. There were occasional sympathetic shorter reviews of his exhibitions during the 1950s, first at the Betty Parsons Gallery, where he showed almost yearly from 1949 to 1954, and then at the Sidney Janis, where he showed in 1955 and 1957. He had shifted to Janis because, as he told Betty Parsons, "I'm doing you no good and you're doing me no good." It was true enough that while Parsons went to great lengths to sell Rothko's works, even taking cuts in her own commissions, he was not flourishing. He had a reputation of distinction, though, and critics such as Thomas B. Hess were attentive whenever a work of his appeared. In 1955 (*Art News*, Summer) Hess observed that "Rothko approaches the classical, Renaissance problem of achieving the elemental serenity of symmetry in a way that avoids the paralyzing boredom perfect symmetry aspires to. Like Bramante or Piero della Francesca, he asks the riddle: 'What is living and stable?' and his answer is a balance of equalities disguised by scale."

162

The fact was that Rothko longed to be understood and he keenly missed searching commentary. As early as 1950 he had told Seitz that writing on art should never be comparative, historical, or analytical, but should record direct responses "in terms of human need." And to Katherine Kuh, he had written in 1954:

> If I must place my trust somewhere, I would invest it in the psyche of sensitive observers who are free of the conventions of understanding. I would have no apprehension about the use they would make of these pictures for the needs of their own spirits. For if there is both need and spirit there is bound to be a real transaction.[74]

Real transactions, Rothko felt, could not occur in conventional criticism, much as he craved understanding. It would not be until his one-man exhibition at the Museum of Modern Art in 1961 that the "meaning of a man's life work" would be pondered by many intelligent writers, and that Rothko could find a measure of satisfaction. The exhibition was organized by Peter Selz, whose enthusiasm made him a willing accomplice to Rothko's eccentric views about how his work should be presented, and what should be represented. Rothko decided to omit work before 1945, and to stress the work of the 1950s. He hung his paintings closely, and, attempting to fulfill the conditions of an intimacy, a space that takes the viewer into it, he worked with very low grade lighting, allowing the light of his canvases to breathe freely. During the exhibition Rothko visited the museum often. He had effectively moved into a new phase, and even his way of thinking about his work had altered. He told Selz, who quoted him in the catalogue, that as he had grown older Shakespeare had come closer to him than Aeschylus because "Shakespeare's tragic concept embodies for me the full range of life from which the artist draws all his tragic materials, including irony; irony becomes a weapon against fate." The injection of irony—a quality relatively absent from most American painting—was a defensive weapon for Rothko, whose general uneasiness in the world grew rapidly as his fame augmented. He was in his late fifties, and he knew that in spite of all the adulation, he would never become a patriarch. Already the "art world" was moving elsewhere, enshrining and thereby leaving behind the old master Abstract Expressionists.

Still, the exhibition drew an unprecedented response, and there were numerous reviews, some showing that their authors had undergone the experiences that Rothko thought could be available in his "tragic" works. Georgine Oeri, a European curator at the Guggenheim Museum, exceptionally sensitive, wrote an appreciation of the exhibition that eloquently spoke for those who were moved by Rothko's canvases. Oeri had been quietly appreciative for several years. With her there had been "a real transaction." She avoided the traps of history and analysis that Rothko so scorned. Her culture made it possible for her to understand what Rothko meant when he spoke of irony as "a sort of self-effacement, a self-examination in which a man can for a moment escape his fate," and when he insisted that an artist makes a decision about "the kind of civilization he tries to present," adding that he was interested only in *this* civilization. Oeri understood the edge on which Rothko balanced so precariously, writing that Rothko's audacity "is not unlike that of Moses who dared to say 'I will now turn aside, and see this great sight, why the bush is not burnt.'" She well understood Rothko's suspicion of critics, writing in her first paragraph of response to the Museum of Modern Art show:[75]

> The audacity of the inspired is never welcome; neither is their testimony. It implies demands upon the world which are resented or rejected all the more because the right or justification to make these demands can never be verified. That which is obvious is not revealing to everybody . . .

Oeri goes on to say that there are good reasons not to propel Rothko's work into "eschatalogical" dimensions, but states what she sees as Rothko's fundamental premise: the confidence in the naturalness of the spirit and in the possibility of its being manifest. She saw that the exhibition as a whole "evoked the sensation of a living organism. All of its parts shared the animation of being in and of the present, participated equally in the composition of a festive and grandly serious place." Like Willem de Kooning, whose response to the show was to speak of a house with many mansions, Oeri understood Rothko's preoccupation with reciprocal spaces and gave one of the best descriptions of how the show looked:

The exhibition galleries were interconnected in such a way that at various crossroads, one faced the huge expanses of the closely hung Rothko canvases, no matter which way one turned. The more they looked different, the more they looked the same and the more they looked the same, the more they looked different.

As Rothko moved from the late 1940s into the 1950s, Oeri remarked that the space "vacated by the early world of phenomena is filled with color which now becomes the evocative instrument," an instrument, she says, that is "impelled to transcend its pictorial quality, to become itself the transforming agent of the substance of which it is made, rendered permeable by the meaning it embodies." Oeri understood a drive in Rothko that would only later become apparent to others when she says of his color, "It is the substance by which that is substantiated which has no substance."

Oeri's opening with the Mosaic allusion would be enough to put Rothko off. Robert Goldwater, however, in his review of the exhibition, won his approval. Goldwater did not attempt to penetrate Rothko's "façade." Rothko was so pleased that he insisted that Goldwater's article be reprinted in the catalogues in Europe when his exhibition traveled in 1962. Goldwater opened his reflections on the exhibition with a bold insight:

> In Mark Rothko's picture the apparent end lies close to the apparent beginning—so close, in fact, or in apparent fact, that they are almost indistinguishable.[76]

He then follows with gratifying (to Rothko) obedience to Rothko's public statements, saying that "Rothko's concern over the years has been the reduction of his vehicle to the unique colored surface which represents nothing and suggests nothing else." Here, Goldwater no doubt missed Rothko's irony. "They say they are façades," he used to say with a malicious chuckle, "so they *are* façades." Goldwater shares Rothko's contempt for "literary fancies" and tries to persuade his readers to forget such "program notes" and to pay attention to the canvases themselves. Yet, to Rothko himself, his colored surfaces, although they "represented nothing" did, contrary to Goldwater's assertion, suggest something else. He never abandoned the will to portray. His paintings, he would maintain in pri-

29. Mark Rothko on his birthday in 1960

vate, were "portraits" of states of the soul, and inevitably were both objects in themselves, and something else. The element of allegory was never foreign to him.

When the British assessed the exhibition, they were rapt. Bryan Robertson, writing in the catalogue for the Whitechapel Gallery (Oct.-Nov. 1961), marveled at the body of work and its emotional power, concluding that "We are left with a presence rather than a specific identity." David Sylvester, writing in *The New Statesman* (Oct. 20, 1968), thought the paintings "the complete fulfillment of Van Gogh's notion of using color to convey man's passion," while Alan Bowness in *The Observer* (Oct. 15, 1961) thought that "Rothko plays Seurat to Pollock's Van Gogh." The painter-critic Andrew Forge was overwhelmed. "When I first saw Rothko's work I felt I had fallen into a dream,"[77] he said, adding that the imperative to go up close was compelling. As the show traveled, there were other strong responses, mostly of awe, and Rothko's reputation as a master was confirmed in capital after capital. And yet he was uneasy. He sometimes doubted greatly. He often fell into depression. He felt not at home in the world, and above all the art world that in the 1960s seemed so promiscuous, so superficial. By the mid-1960s, Philip Guston reported, "Rothko too thought that the smoke that existed ten years ago was a false situation; this is the *real* situation."[78] In this "real" situation of isolation, brought on by fame and the waning of New York School influence, Rothko continued to work for a time in the moody vein established in the two mural cycles but felt increasingly nervous about his direction. His craving to "control the situation" had grown more insistent, while "the situation" had become less and less clear.

10

The controversy over the real meaning of Rothko's work (meta-physical or physical?) would be rendered pointless when, in 1964, he eagerly seized the opportunity to create a Catholic chapel in Houston, Texas. Christian themes were not foreign to him. He had painted a Last Supper in his Greek manner in the late 1930s, and a Crucifixion and Gethsemane. In the past he had tended to treat Biblical themes as part of the great mythological reserve Frazer had provided. By 1964 he unequivocally saw "human need" in terms of spirit. All the questions hovering around the meaning of his work were to be definitively answered in this undertaking. He felt increasingly uneasy in the world of interpreted things. He had had many honors but they provided transient satisfactions. While he accepted certain homages—such as the invitation to John F. Kennedy's inauguration and a state dinner for artists a few months later—he felt increasingly isolated and held himself apart. Occasionally he would indicate his disaffection in spirited tirades against the art world, sometimes to people he did not know well. Or, he would pour out his disappointments, angrily denouncing insensitive critics. He objected to those who were busy tracking down his Surrealist sources, telling Andrew Forge in 1963 that he had never swallowed the vulgarities of psychoanalytic jargon, and that his painting was far removed from the precincts of the unconscious with its supposed archetypes. In addition, he was increasingly appalled by the cultural and political situation and, like many other artists, looked upon the pumped-up promotion campaigns now attending the arts as inimical to the purposes of art. So, when the de Menils, who had admired his Seagram murals during a studio visit in 1962, came later with their offer of an entire chapel, Rothko was keenly receptive. John and Dominique de Menil were sympathetic patrons. They had a long history of friendship with artists and an exceptionally fine collection of modern art. They had shown understanding when Rothko had tried to explain why he finally rejected the Seagram commission. The de Menils were willing to invest complete faith in whatever Rothko would do. Philip Johnson drew up the original floor plan as an octagonal shape. This "pleased Rothko who had a special liking for the Torcello baptistry and church," Mrs. de Menil wrote.[79] As soon as plans had been drawn Rothko set to work in his new carriage-house studio on East 69th street, constructing a full-scale model of a segment of the chapel, simulating as closely as possible the ultimate structure.

Rothko had been given the commission in the late spring of 1964. In the summer he had taken his family to a rented cottage in Amagansett, Long Island, overlooking Gardiner's Bay, where he mused on his project. The commission seemed to bring a great release. During a visit I saw Rothko on his porch watching tenderly as his baby son Christopher clapped his hands while Schubert's Trout Quintet filled the air. He spoke distractedly of small domestic things during a lunch of hamburgers. After, we walked on the beach, and he became unusually animated. He talked of the new studio and the commission. His idea was to "make East and West merge in an octagonal chapel." This was an old dream. He had often spoken of reading the Patristic writers during his youth—a part of his self-education of which he was proud, and which seemed very odd to his earlier acquaintances. But Rothko's attraction to Origen and the other early Christian writers was completely in keeping with his temperament. What he liked, he said, was the "ballet of their thoughts." He said that in them, everything "went toward ladders." It was not unlikely that he was attracted to the Patristic writers because they were in rebellion against fixed traditions, and particularly against the empty rhetoric they discerned in late Greek thought. The fathers of the church were careful to mention as often as possible (at least in one aspect of their works) their contempt for "style" and their belief in elementary simplicity. The early fathers of the church spoke in a period of transition in which individual views were still valid and the church was not yet rigidly institutionalized. It was, as Werner Jaeger sees it, a historical encounter between two worlds, the Greek and the early Christian.[80] The Greek influence is rendered in the conception of the universe as an organic being. All its aspects are functional, as in the human organism; all the parts "breathe together," as Clement of Alexandria wrote. Moreover, the early fathers carried a humanistic tradition from Greek culture, often using the word *morphosis,* meaning "the formation of man," and not the theological terminology that dominated later Christian thought.

Rothko's harking back to the Patristic thinkers was a necessary move for him. Every artist seeks or creates a tradition within which he can feel unique. Rothko had to invent a tradition, or a fiction of a tradition, because it is only in the contents of a life of the mind, that includes everyplace it has wandered, that an artist can find his style. Sometimes he must abandon his present in order to find it again and test it against the universal human table of contents he has carried for so many years. Rothko instinctively sought another context. He felt he was hemmed in by contemporary clichés. At times, he felt he had reached an impasse in his enterprise. Now he could set out again, gathering all he knew of human existence, in order to express its potential to transcend. He had traveled from self-expression to the grandly tragic as suggested by Nietzsche; from symbolism to silence; from Fra Angelico to the early Christians. Shuttling backward and forward in human history, spirituality was his goal.

In the "ballet" of the thoughts of the Greek fathers of the church, there were many endearing allusions to the visible beauties of life in the world; many simple descriptions of life in monastic retreats. There was even jesting and teasing in letters exchanged. The view of the world expressed by Origen and generally held by the early fathers was that of a whole organism, "some huge and immense animal which is held together by the power and reason of God as by one soul." Creation, according to Gregory of Nazianzus, is "a system and compound of earth and sky and all that is in them, an admirable creation indeed when we look at the beautiful form of every part, but yet more worthy of admiration when we consider the harmony and unison of the whole, and how each part fits with every other in fair order, and all with the whole, tending to the perfect completion of the world as a unit."[81] Reading such passages, Rothko primed himself for his task. In Gregory of Nyssa, Rothko could warm to the dream of many artists, symbolized in the ladder, known so well through the terse, mysterious passage in the Old Testament where Jacob lies dreaming. Ladders, wherever they appear in art, are instinct with symbolism that soars to spiritual realms. Gregory of Nyssa wrote that the man of half-formed intelligence, when he observes an object bathed in the glow of apparent beauty, thinks that the object in its essence is beautiful and goes no deeper. A more developed mind will see outward beauty as the ladder by which he climbs to that intellectual beauty from which all other beauties derive their existence, and their names, in proportion to their share in it.

Perhaps it was Origen's tendency to allegory, and his method of threefold interpretation that attracted Rothko. Origen described the three approaches to the scriptures as literal, ethical, and allegorical or, literal, historical, and spiritual, in another translation. Rothko had had his literal phase; had been preoccupied with ethical interpretations; and would, finally, arrive at the ultimate spiritual allegory in the chapel. The whole period in which the Greek theologians put forward their views attracted Rothko. Instinctively he was drawn to the cryptic, the hermetic. He liked the old baptistry in Torcello and Byzantine structures in general. In them, there is nothing "outer." The outer walls of Byzantine chapels are blank and impenetrable. Unadorned, they are merely the keepers of the treasure within. Rothko had seen Ravenna. The little cruciform chapel of Galla Placidia, with its nocturnal light created entirely within the tesserae crowning its dome, and its tiny windows of alabaster, lingered perhaps in his memory, for there are few monuments in history with a space as complete and as controlled by the artist whose signature is in the unique light he has produced. An inscription in the archiepiscopal chapel in Ravenna (certainly it would be a discovery of moment for Rothko) reads: "The light is either born here or, imprisoned, reigns here in freedom" (*Aut lux nata aut capta hic libera regnat*). The central apse in the basilica in Torcello also stimulated Rothko's vision of spiritual light. There is a rare apparition of an elongated madonna and child, standing on a symbolic, disembodied small rectangular platform against a vast absence created in gold. The totally abstract gold ground of the curving wall of the apse gives an immense sense of transcendence. For the worshipers of the late antique period, the churches were *ho topos*, "the place" in which, as Peter R. L. Brown has written, "it was possible to share for a moment in the eternal repose of the saints in paradise. Light seems trapped in the churches. The blaze of lamps and gold mosaic recapture the first moment of Creation: 'Dark chaos is fled away.' "[82] For Rothko, *ho topos* had been an obsession for many years, and the flight from dark chaos an ideal of long standing. As he worked he seemed intent on making "the place" where, in its grand abstract realm, necessity is fled along with chaos.

When Rothko returned from Long Island in the fall of 1964 he began working in earnest. The huge studio with its central skylight shielded by parachute silk became a place as he imagined it, different day after day, and re-made each day. As in his old gymnasium, there was little to distract him. A bed, a couple of canvas captain's chairs and hi-fi equipment were the only amenities visible. A series of young assistants came to help him build the huge stretchers which in themselves, even before they carried their canvas surfaces, were of crucial importance to Rothko's imagined "jointed scheme." There were times when Rothko sat for hours in his canvas chair, contemplating the shape and size of the empty stretchers. His assistants would be called upon to change one or another as Rothko sought the absolute solution to his fourteen-sided scheme. Eventually the job of stretching the canvas replaced that of contemplating. That would take nearly a month, so careful was he that surfaces be exactly even and permanently stretched. Then Rothko would begin his painstaking procedure of preparing the ground—a procedure recommended by the old masters in which he boiled rabbit's glue and mixed dry pigment together with oil and a little turpentine. These grounds, composed very largely of Alizarin crimson and black, were laid on swiftly by his assistants, and had to be smooth but not monotonous—a job that kept his assistants sweating. In these decisions Rothko was slow but assured. Once the canvases had their "plum" grounds, they would be aligned in the replica of the chapel and Rothko would again sit and contemplate them. Finally, as Roy Edwards, one of his assistants recounts, he would decide where to place a black rectangle and the assistant would dust it in with charcoal.[83] These interior shapes were then contemplated for days, singly and in relation to others. Rothko would sometimes decide to change the distance of the dominant shape from the border by a quarter of an inch, or, if he had already painted the dark interior, he would put the canvas away and substitute another. Months went by with this elaborate procedure, with Rothko inching toward his vision.

Rothko's choice of two basic colors—black and red—and their light-reflecting variants helped to define his image (for it was to be a single image finally). The degrees of light available in these color juxtapositions were calculated broadly, for it was an *effect* he sought and not a mechanical scheme. He was enough of a Nietzschean to be wary of systems. Nietzsche had nothing but contempt for systems that were airtight, writing,

> The will to a system: in a philosophy, morally speaking, a subtle corruption, a disease of the character; amorally speaking, his will to pose as more stupid than he is, more stupid, that means: stronger, simpler, more commanding, less educated, more masterful, more tyrannical.

Rothko had come to think of his chapel as a matter of finding the right proportions in an almost mystical way, like the alchemist. Proportion with its manifold implications: the right proportion, as the Greeks had suggested to him in his youth, tempers a man's life; the right proportion of pigment to oil brings him his light; the right proportion of thing to infinite space defines a man's stance in a world only he can construe. Morton Feldman understood as a fellow artist can understand when, of all his recollections of encounters with Rothko, he chose to talk about a visit to the Metropolitan Museum in which Rothko made his way to a small room of Greco-Roman sculpture:

Rothko always followed through his reaction to something that would catch his attention with a brief, reflective commentary. I remember his absorption one afternoon with the Greco-Roman sculpture: "How simple it would be if we all used the same dimension—the way these sculptures here resemble each other in height, stance, and the distance between one foot and the other."

Feldman's comment is astute: "Rothko was leaning toward a possible *answer* in a more subliminal mathematics of his own work."[84] Such "subliminal mathematics" has to do with one of the most perplexing and mysterious problems in the visual arts—the problem of scale. Every art student knows that scale is not the same as size. And most art students have been accustomed to thinking about scale as a problem of relative proportions. But Rothko, and some of his colleagues in the New York School had had a glimpse of another way of creating scale—a dazzling vision that not even the most intelligent among the painters could quite pin down. He envisioned his task in the chapel as a clarification of this eternal painterly problem. He would, with the right proportions of color and shape, substantiate, as Oeri wrote, that which has no substance.

There were often visitors to the studio and Rothko would sit, contemplating his work in progress, awaiting responses, although no one could gauge how much he was dependent on response. When old painting friends such as de Kooning or Motherwell or Guston visited, there was talk about "the situation," and sometimes about the murals. Motherwell visited early in 1967 and came away with a profound respect. "They are truly religious," he commented, adding that Rothko had said that he had become "a master of proportion." Motherwell quoted Rothko as saying that in the beginning he had thought of them as pictures. But then, he considered that people praying would not want to be distracted by pictures. They wanted an ambiance.[85] Of course an ambiance is what Rothko wanted, composed of rhythm, proportion, and transcending emotion—rhythm in the sense that Hölderlin sang of it,

> All is rhythm; the entire destiny of man is a single
> celestial rhythm, just as the work of art
> is a unique rhythm.

Rothko, when he made the decision to leave certain of his murals bereft of image was certainly thinking of a unique rhythm. He was shaping the "absent," as Mallarmé was shaping it in his singular final work "Un Coup de dés." Like Mallarmé he had gone from a technique of metamorphosis to something nearer in meaning, to transfiguration. Proportions seemed to be the key. In the context of Rothko's last works, it is possible to understand what he admired in Mondrian (for he *did* admire Mondrian, and was proud when Mondrian's disciple Fritz Glarner told him that he was closer to Mondrian than anyone else. He even lectured on "the sensuousness of Mondrian").

Rothko had said that a painter's work would move toward clarity. When in January 1965 he wrote a careful memorial speech for his old friend Milton Avery—a speech for which there were several well-worked drafts—he was himself at work on the Houston murals. His thoughts turned to Avery's walls "covered with an endless and changing array of poetry and light." Avery, he said, "had that inner power in which gentleness and silence proved more audible and poignant." Most important, Avery had nothing "tentative" about him, he had "naturalness," "exactness," and the "inevitable completeness" that Rothko posited as the highest values. Such thoughts had no locus in the contemporary art world from which Rothko felt so alienated. Rather, he would have had to turn back to find another climate of thought. Like Stephen Dedalus, he found himself in another era in which the question of *claritas* seemed urgent. Joyce's need to discuss the old scholastic notions of aesthetics sprang from a similar disaffection with the art of his time; the same need to find a new approach by rejecting the clamor of the day's encounters. In "Portrait of the Artist" Stephen wanders about the streets of Dublin discoursing on the Thomist aesthetic, trying to find his way among the three things Aquinas has said are essential to beauty: wholeness, or *integritas;* harmony, or *consonantia,* and radiance, or *claritas.* Radiance Joyce decided was equivalent to the scholastic notion of *quidditas,* or the *whatness* of a thing, without which a work of art cannot be realized. Joyce considered the highest form of art to be dramatic, and even in this early work, envisioned his life's task. The dramatic form is reached, he said,

> when vitality which has flowed and eddied round each person fills every person with such vital force that he or she assumes a proper and intangible esthetic life. . . . The esthetic image in

the dramatic form is life purified in and reprojected from the human imagination. . . . The artist, like the God of creation, remains within or behind or beyond or above his handiwork, invisible, refined out of existence, indifferent, paring his fingernails . . .

This medieval frame for his thoughts suited Joyce's instincts as an artist, and were equally suitable to Rothko's final enterprise. His monastic intentions are undeniable in this last expression of universal silence—intentions as ascetic as Fra Angelico's in the late works for his monastic brothers. This final work would be "the place" where the artist, as Rothko had said a few years earlier, would surpass his "self." He took his task absolutely seriously, as his letter to John de Menil (Jan. 1, 1966) suggests:

> The magnitude on every level of experience and meaning, of the task in which you have involved me exceeds all my preconceptions. And it is teaching me to extend myself beyond what I thought was possible for me. For this I thank you.

The will to sink into a whole, the dissolution into a universe that, as Nietzsche and then Joyce had imagined, would absorb the individuality of the artist, had always been latent in Rothko. Wells, caves, retreats are implicit in the depths of certain of his works where the light has been dimmed almost to extinction. Why had he painstakingly developed a technique of overlaying colors until his surfaces were velvety as poems of the night? Sometimes in his works of the late 1950s he would allow hints of striations in the middle sections of his paintings—bars of indeterminate color that suggest the river flow, eternal flux. The elements had breathed in his works, and for the chapel, he would still maintain their pulsation under the final surfaces of his panels. Light was meant to flow from one to the other, unimpeded by detail. The scheme for the chapel would fulfill his vision, expressed some fifteen years earlier, when he told Seitz that antitheses are neither synthesized nor neutralized in his work but held in a confronted unity which is a momentary stasis.

Such stasis is not foreign to the religious experience. The question has often been raised: was Rothko religious? I don't think he was religious in any conventional sense. More likely he was religious in the way Matisse was religious when he undertook the Vence chapel. Matisse said, after its completion, in his letter to Bishop Rémond that this work represented the result of his entire life and that it issued from a life consecrated to the search for truth. In his notes for the chapel booklet, he said that the chapel had afforded him the possibility of realizing all his life's researches "by uniting them," and that the chapel was the flower of an "enormous, sincere and difficult effort."[86] To the degree that the magnitude of Rothko's task on every level exceeded all his preconceptions, as Rothko wrote, he put himself in what might be called a psychological condition of religiousness. This was not as difficult for him as might be thought. This Russian, who had consumed his Dostoyevsky and had always charged himself with ethical conundrums, could imagine a place in which there could be solace for the secular. His chapel would not be an experience of private meditation such as those who contemplate mandalas undergo, but an experience of the summum of a spiritual man's life's work, where all his researches, in Matisse's terms, could be united. The expression of faith had to be a faith in the Existentialist idea of intersubjectivity—the only faith left to modern man. The highest praise in Rothko's vocabulary, as Motherwell remarked, was to call someone a "human being," that is, a person who feels.

Yet, Rothko was a Jew, had learned to pray in Hebrew, and known the humiliations of Jewishness. He always said he would never have worked for a synagogue. He did not explain, but one could speculate that for him, the chapel was already a distancing, an opportunity to stand back and generalize his deepest feeling about existence. He could not have stepped back in a synagogue in which so many conflicting emotions resided for the modern Jew. He did think of himself as a Jew, as is apparent in a dramatic encounter ingenuously described by the German art historian Werner Haftmann. Haftmann visited Rothko in 1959 to ask him to participate in "Documenta," the large international exhibition. Rothko refused saying that "as a Jew, he had no intention of exhibiting his works in Germany, a country that had committed so many crimes against Jewry."[87] After further conversation, Haftmann reported, Rothko looked at him squarely and said that if he "could manage to have even a very small chapel of expiation erected in memory of Jewish victims, he would paint this without any fee—even in Germany, which he hated so much. He then said it need only be a tent." This passionate expression of Rothko's deep personal response to the Holocaust led Haftmann to fanciful conclusions concerning the nature of Rothko's enterprise that, had Rothko lived to read them, would probably have appalled him. Haftmann seized upon the "tent" allusion and wove a fantasy in which the "swaying" quality he detected in certain of Rothko's largest pictures took on the character of curtains which in turn evoked old Jewish metaphors of the temple curtains in front of the Holy of Holies. This led Haftmann to assume that the source of all of Rothko's work lay in his "Eastern Jewish humanity" and that Jewry, which remained amorphous for more than two thousand years, had found its own pictorial expression.

This interpretation denies one of Rothko's most impelling arguments: that he was a 20th-century man nurtured on Nietzsche and well aware of the futility of ancient gestures. Like the young Malevich, whose spirit his most nearly resembles, Rothko sought a godless expression of godliness. He too had emotional responses that, as Malevich said, led to the desert of pure feeling. The kind of universe Malevich had in mind was boundless, yet encompassable or "felt" by the imagination which could intuit its very boundlessness. Malevich built for himself an objectless world. Swept clean of centuries of painterly clutter, Malevich's world, expressed for him first in a black square on a white ground, was secure from the ravages of daily life. It was "elsewhere." The diction of his manifestos indicates his fervent will to translate himself into another climate—one of great clear spaces and pure light. His written language is ecstatic, and he envisioned a plastic language equal to his visions. This plastic language of "weights" and "movements," rather than forms and colors, was posited to be read by kindred spirits in whom Malevich had confidence. Eventually his project was confirmed: generations of artists were at home with this objectless language of pure feeling. It became one of the available idioms for Rothko's generation. Not only could they wield its philosophical principles, but they could, as did Rothko, see Byzantine and Trecento art through its lens. If

Rothko insisted that he was a "materialist" and a 20th-century man, he meant that he was forsaken in the world, as all humans are, just as Nietzsche said man would be forsaken until, by his own will, he would become an overman, in control of his personal destiny. The thoughts of a man of the cloth, Father Couturier, who had engaged many modern artists in church decoration, are closer to Rothko's situation than those of the art historian: Father Couturier thought that "Genius doesn't give faith, but there is between mystical inspiration and that of heroes and great artists a profound analogy."

As Rothko sat in his captain's chair, contemplating his masterwork month after month, he had diverse thoughts, some of which he voiced to friends. Early in 1966 he looked at one of his mural sequences and said, "I'm only interested in precision now." He stressed his new quest for proportion when he said "even when I used to use three colors, say purple, yellow and red, if I saw them in relativity, I had to do something else." In the fall, after he had returned from an important sojourn in Europe, he mused, "You never know where your work will take you," adding that in the Houston cycle he was interested neither in symmetry or asymmetry, but only in proportions and shapes. In early spring of 1967, when a mysterious melancholy had overtaken him, Rothko was contemplating one of the triptychs—the center panel an obscure color of red-brown-gray mixture, with oxblood borders; the two side panels uniform oxblood—and remarked, "I wanted to paint both the finite and the infinite." This reminded him of the work of Ad Reinhardt, who had resolutely eliminated both color and individual form from his recent work. "The difference between me and Reinhardt is that he's a mystic. By that I mean that his paintings are immaterial. Mine are *here*. Materially. The surfaces, the work of the brush and so on. His are untouchable." His thoughts then shifted to the morning's crisis in Israel, the so-called war of attrition, and he asked, as he often asked in his last years, "Will the world last another decade?"[88] The somberness that most people felt in the studio; the feeling of solitariness; the kind of hushed space in which Rothko worked in his last years made for an atmosphere that could be perceived as sacred, or at least, associated with the interiors of sacred places such as Byzantine churches. However, Rothko's vision, at least in its

30. In Rome, 1966: Rothko, the painter Carlo Battaglia,
Rothko's son Christopher, and his wife Mell

religious dimension, if there were one, was not bounded by his early Jewish formation, or by his admiration for any other culture, or by any special feeling for Catholicism. If anything, the light in the Houston murals called to mind Pascal's "Deus absconditus," a far more troubled and modern conception of the fled deity.

As Rothko carried his work along from month to month and year to year, he struggled to refine his conception, and his mood darkened. In a kind of aesthetic despair (whose sources he had long ago examined in the light of Kierkegaard) he worked out many schema in which differences became increasingly minute. By the spring of 1966 he felt he had settled on a satisfactory but perhaps not final scheme and decided to take a break by making a journey to Europe. Once there his depression lifted, and, especially in Rome, he set out to re-examine places that had moved him on his previous trips. Comfortably settled in the Battaglias' spacious apartment in the Palazzo Cenci, Rothko spent hours lounging on the leather couch, listening to *The Magic Flute* and *The Abduction from the Seraglio*, abstracted, removed, perhaps reviewing his project. At times he would stand at the window overlooking the piazza below in which a fountain played softly, singing to himself. In late afternoon he saw swallows swiftly descending to the basin of the fountain. He found a kind of peace in Rome that was to be later translated in the Houston chapel.

He also found warm responses and good conversation, and, as was often the case with Rothko, he revealed himself more fully to those with whom he had not had long and close relationships. In Rome he revisited Argan, where he discussed Fra Angelico again, and he spent time with the Scialojas. With Carlo Battaglia, he visited the oldest churches in Rome, peering closely at mosaics and frescoes from the earliest Christian period. They made a trip to Arezzo as well, where Rothko was dismayed. The Pieros were too much like illustrations he said, comic strips even. He did not like the Cinquencento and preferred the more conceptual mysteries of the early Renaissance. He tended to look at painting as a message, Toti Scialoja noted, recalling a visit to his own studio where Rothko looked seriously at one of Scialoja's abstractions with vertical forms and finally said, "I can understand that two are man and woman, three are man, woman and child, but five are nothing." Rothko could not like Piero, Scialoja felt, because Rothko "voyait avec des yeux symboliques."

From Rome the Rothko family traveled to France, Belgium, and Holland, concluding their journey in England. England had long been a beacon for Rothko, who had found a specially cordial reception during his first visit in the 1950s. The British painters had responded with exceptional fervor to their first exposure to his work in 1956 when the traveling exhibition *Modern Art in the U.S.A.* was shown in London. On each subsequent occasion when Rothko's work appeared in London (in the *The New American Painting* at the Tate in 1959, and in his one-man exhibition at the Whitechapel in 1961) enthusiasm mounted. Critics in Britain were generally reverent and often wrote lengthy articles that must have gratified Rothko, who always felt the critics in the United States were short of breath. The adulation of younger painters and their unreserved praise warmed him. A glimpse of what Rothko meant to that first postwar generation is caught in a letter to the *Times* of London written by William Scott and Paul Feiler after Rothko's death. They spoke of "his great human qualities" and pointed out that "when his work first appeared in England there was an immediate response on the part of a number of younger generation painters who recognized that here was a vital contribution made to the heritage of European landscape tradition through colour and light." They commented on his love for the English countryside and observed that "his knowledge of English literature was remarkable." They concluded, "Perhaps much of what happens today stems from that aesthetic shot which Mark Rothko gave us in 1956."

Rothko found satisfaction in the unreserved admiration tendered him during the course of the exhibition at the Whitechapel Gallery. The director of the gallery, Bryan Robertson, was sensitive to Rothko's needs, and had a strong feeling for the work. He made sure to report to Rothko that Henry Moore had made a special journey to London to see the show; had visited many times alone, and had said that Rothko's paintings were "his most revelatory experience in modern painting since his youthful discovery of Cézanne, Picasso, Matisse and the Cubists." Robertson added that the same was true of Sir Kenneth Clark and many others. "Rothko really loved England, and his affection for us and our foibles was so candid and unaffected that it took some time to realize that he was not joking, for even the inconveniences of English life for Rothko had a certain charm."[89] During his last visit Rothko conferred with Sir Norman Reid, the director of the Tate, and Andrew Forge, a trustee, over a long lunch, discussing the eventual Rothko room. Forge remembers Rothko's excited diatribe against Piero della Francesca, which led into a long speech against the idea of figuration in painting. "He spoke with a real sort of iconoclastic anger," Forge recalls. Both Reid and Forge sensed an uneasiness, probably related to Rothko's feeling that the world had turned to other things (things such as figurative art in its pop version, for instance). Reid mentions that Rothko asked again and again if he thought the young painters would be interested in the projected Rothko room, and Forge sensed that Rothko felt that he was, to the younger British connoisseurs, passé (which, as Forge points out, was probably true. By that time England had had major exhibitions of work by Robert Rauschenberg, Jasper Johns, and others in the new vanguard). To counteract Rothko's depression, Forge and the critic David Sylvester organized an evening with British artists who in fact admired Rothko greatly. But Rothko was not to be easily consoled.

When Rothko returned to his studio he had clarified his conception still further and set about making adjustments. His conflict with Philip Johnson over the eventual lighting of the chapel worried him. Rothko rejected Johnson's idea of a truncated pyramid that would allow light to diffuse the walls evenly, preferring to reproduce in Houston the skylight of his own studio. Mrs. de Menil recalled that "his love for familiar surroundings was such that he wanted also to have the same cement floor, and the same kind of walls . . . He liked irregularities, accidents. He liked ancient buildings with odd shapes, grown from 'the weaknesses and follies of men.' "[90] The conflict was only resolved when Johnson withdrew and the architects Howard Barnstone and Eugene Aubry took over, visiting Rothko and attending to his desires. (Johnson maintained later—and he was probably right—that Rothko had made a mistake about the lighting.)

The final scheme was to be fourteen canvases distributed symmetrically. In Mrs. de Menil's description:

> In the apse; a triptych of monochrome paintings fifteen feet high. To the right and left: two triptychs eleven feet high which had black fields. All the other paintings are again fifteen feet high. The four paintings on the four small sides of the octagon are monochromes, and almost eleven feet wide. The one at the entrance, only nine feet wide has again a black field.

In the actual chapel, when the paintings were installed after his death, the effects were disappointing. Rothko had been right to cherish "the place" that he had made in his own studio. The new place, with its concrete and steel, was, despite best intentions, far too perfect to be perfect for Rothko's vision. The problem of the searing Texas light could not be solved and it was only at certain moments late in the day that the chapel took on a semblance of the sacred aura so many had felt in Rothko's own studio. All the same, countless visitors to the Texas chapel (which had eventually been transformed into an ecumenical rather than Catholic chapel) experienced a kind of secular solace. Stanley Kunitz recalls a whole history of feelings. First, almost depression—they seemed to have such a bleakness of tonality. Later he grew to see the paintings in the changing light where he caught the sense of tranquility, and saw "all kinds of fluctuations." When he read his own poems in the chapel, finally, he was immensely moved.[91] This slow discovery by a poet is probably the best tribute to Rothko's enterprise. Another response by the art historian Robert Rosenblum stresses that the paintings evoke a traditional religious format:

> On three of the chapel's eight walls—the central, apse-like wall and the facing side walls—Rothko provided variations on the triptych shape, with the central panel alternately raised or level with the side panels. Yet these triptychs, in turn, are set into opposition with single panels, which are first seen as occupying a lesser role in the four angle walls but which then rise to the major role of finality and resolution in the fifth single panel which, different in color, tone and proportions, occupies the entrance wall, facing, as if in response, the triptych in the apse. It is as if the entire current of Western religious art were finally devoid of its narrative complexities and corporeal imagery, leav-

ing us with the dark, compelling presences that pose an ultimate choice between everything and nothing . . . the very lack of overt religious content here may make Rothko's surrogate icons and altarpieces, experienced in a nondenominational chapel, all the more potent in their evocation of the transcendental . . .[92]

Not all responses were equally sympathetic. Brian O'Doherty, who had harbored feelings of distaste for Rothko's occasional "schmaltzy sentimentality," questioned Rothko's position, reporting that Rothko had boasted that he could have fulfilled the commission with blank canvases "and made it work." Rothko's acute anxiety, O'Doherty declared, was "based on a suspicion that his most unsympathetic viewers shared—was there anything there at all?"[93] Such skepticism certainly lingered in the minds of many whose expectations were challenged by the final austerity of Rothko's scheme. Those who could believe were those who had a touch of the same aspiration, such as the composer Morton Feldman, who has described his music as "between Time and Space," saying that his compositions are not really compositions:" "One might call them time canvases in which I more or less prime the canvas with an overall hue of music." Commissioned in 1971 by the Menil Foundation to write a work in memory of Rothko, Feldman composed *The Rothko Chapel*. In his program notes, Feldman wrote:

> Like the chapel, the music is conceived in an ecumenical spirit. I think of it as a 'secular service.' I tried to create a music that walks the thin line between the abstraction of all art and the emotional longing that characterizes what it is to be 'human.' The chorus symbolizes art's abstractness; the solo viola, the need for human expression. It is only at the end of the work that I think of Rothko and his own love for melody. Here, I collage a Hebrewesque melody which I wrote thirty years ago—at sixteen.[94]

11

In 1958 Rothko had listed as one of the ingredients of a work of art the presence of irony—the "modern" ingredient. Greek tragedy didn't need it he said, but Shakespeare did. By his definition—that irony is a form of self-effacement and self-examination in which a man can for a moment escape his fate—the works of his last years permitted him few moments of escape. He was a natural heritor of a modern grief initiated long before, when doubt assailed the searchers. Pascal's *deus absconditus* was not the same God hidden in the Old Testament who was nonetheless there. Pascal's God eluded him endlessly. Faith would become a negative necessity, like, perhaps, space. The pain of the absence of faith and the great reservoir of doubt became the leitmotifs of the great minds of the 19th century.

Many modern artists had experienced a similar chill when confronting the Nothingness that had first become a motif in the 19th century. The situation of the modern artist was precisely that he was invested with the power to illuminate and inspire, but could not, as artists had before, assume the presence of a prime mover. Rothko was an Existentialist, finally, because he felt he had been as Heidegger said, "thrown into the world" and each time he had moments of ekstasis, he was thrust back into the human condition. O'Doherty was perhaps right when he suggested that Rothko doubted. But it was a high order of doubt. It had to do with the problem of possession: few modern painters can sustain the feeling that they have truly possessed their work, or experience. For Rothko and so many others, it was a matter of trying again, each time as anxiously as the last—anxious about what might be, what could be revealed. Rothko had maintained that the human enterprise in art would be to deal with the great themes of birth, dissolution, and death. "Tragic art, romantic art deals with the fact that a man is born to die."

The tranquil melancholy that prevails in the Houston chapel had not come upon Rothko suddenly. For a long time he had been a sad man. Even when he had achieved international renown on an unprecedented scale and could buy a large town house in an elegant neighborhood, he was perceived by friends as a man subject to serious depression. The quality of Rothko's sadness was understood by the Italian writer Gabriella Drudi, who visited New York in 1960 just after Rothko had acquired the grand house on East 95th Street:

> We went together to visit the house. There was happiness. The festiveness of the poor when they finally know possession. We climbed to the third floor where the studio would be; then to Tofy's room, and ended up sitting in the little courtyard—a bit melancholy as is every place that has been taken over by new inhabitants, still intruders. Mark had become taciturn and grim—this happened, I think, each time that he felt drawn into existence and not art. He seemed very tired. He said to Mell: you remember when I used to pass my days at the Museum of Modern Art looking at Matisse's *Red Studio*? You asked: why always that and only that picture? You thought I was wasting my time. But this house you owe to Matisse's *Red Studio*. And from those months and that looking every day all of my painting was born.[95]

Drudi observes that Rothko had said it not with the pride of one who had known how to "see" where he looked, nor with the firmness of one who tells about himself: "It was more the sad accent of the memory of a love; of the birth of a love." The sadness, the nostalgia for the birth of a love, dominated Rothko's last years. After the chapel paintings had been completed, Rothko was left with a great void. He had difficulties returning to the single easel painting. He felt displaced. Gradually he found a way again and produced several large and magnificent "afterstorms," as Rilke might have said. His working rhythm finally restored, Rothko was in somewhat better spirits when suddenly, in April 1968, he was struck down by a serious aneurysm. The gravity of the attack did not escape Rothko, whose brooding on death was habitual anyway. When he recovered enough to go with his family to Provincetown, he was in a black mood. All the same, he began a new series of works on paper, dark foreboding works of a sinking heart; blacks over purples and blacks over browns with an unaccustomed decisive line separating two rectangular areas in which, as he said, "the dark is always at the top." He himself was startled by these works and asked if it were agony or persuasiveness they represented. "All the crosses we load on our own shoulders," he wryly remarked in the fall of 1968, "when the world settles for things *without* crosses."[96] He felt increasingly remote from the preoccupations of the art world and from artists of the new generations. "We live at a point when the foremen are making the patterns," he commented, and then harked back to his incessant question: will the world last another decade?

Many old questions troubled him. In February 1969, I visited the studio. Highly nervous, thin, restless, Rothko chainsmoked and talked intermittently. Literature and music, he said, were his base. He was never really "connected" with painting, as he started painting only late. His material is his "inner life," his "inner experience." He has nothing to do with painting today, but rather is a Renaissance painter. (I was reminded that in the 1958 lecture he had firmly declared that in the great artistic epochs "Men with their minds produced a view of a world, transforming our vision of things.") With their minds . . . This was Rothko's vision of the Renaissance artist; his ideal which, in his isolation after his brush with death, he felt had been abandoned by all the world. The brooding and often harsh character of the many large paintings on paper he showed me that day could only be seen in the light of a man's sinking heart. The dark, he had said with unintentional symbolism, is always at the top. To Stamos, Morton Levine, Bernard Reis, and a few other close friends, he sometimes spoke of his aesthetic despair and the hollowness of his fame. He was convinced that on the whole he had never been properly understood.

There were many paintings from the last two years of Rothko's life. Some reverted to his older vision, but most were, he felt, new departures for perhaps another destination (although in the end Matisse was right, the destination is always the same). In some, he initiated the glaring white border, the blankness severely emphasized by the perfectly trued angles—with none of the "weaknesses and follies of men" at the outer limits. Only cold white light. The interiors were loosely painted: an area of dark above, a very fine line of light, and the fogged white-grays inflected at times with pinkish or bluish or ocherish hues below. Sometimes, during these last years, there were paintings in oil, and in them, Rothko occasionally used only gradations of black invoking his magical sheens. Like his cherished Fra Angelico, Rothko at the end of his life had no need of a range of colors. There was only one kind of light. The paintings with the brownish-grayish tonalities, done mostly in the winter of 1969, may well have been related to an event that had temporarily lightened Rothko's mood. In the winter of 1969, UNESCO headquarters in Paris had approached him about "doing" a room which would also have sculptures by Giacometti: He had long respected Giacometti whose work he saw as "tragic" in the lofty sense with which he always endowed that word. The thought of inhabiting, with his canvases, the same space as Giacometti's sculptures, was behind the series of brown-above-gray paintings on paper (meant to be mounted on canvas and stretchers). As Motherwell remarks, the colors were "not unlike the colors Giacometti himself used in his own figure paintings." In his conversation with Motherwell, Rothko referred to his new group of paintings as "a different world from myself"—one in which, as several other friends point out, he was not quite at ease.[97] Michel Butor wrote of earlier dark works, ". . . one of the most remarkable of Rothko's triumphs is to have made a kind of black light *shine*."[98] Now, the shine of black light and the shine of substantiated white light would be equivalent in these enigmatic last paintings. Butor refers to the conventional three areas in many of Rothko's earlier paintings as rungs of a Jacob's ladder. This Jacob's ladder was never literal, as the ladder of Jacob himself was not. The movement of the last paintings is equivocal. The resolute horizontality, stopped only by the white bands at the edges, seemingly denies an upward progression. The weighted darks at the summit are invested with deepening emotion. Long before, the French commentators had determined that, despite Rothko's disclaimers, his work was the work of a mystic. Philippe Sollers had written that "Rothko's interiorized, fiery furnace; his patently sacred enclosure where everything indicates a spiritual ambition . . . are opposed to the profane and radically rational space of Mondrian."[99] The painter Stamos was sensitive to the implicit tragedy in these last works and saw them as Goyesque.

Robert Goldwater, who had been authorized by Rothko to write a serious study of his life's work, edged closer to the kind of romantic reading he had once forsworn. Goldwater had spent time with Rothko during the difficult last period and reported that Rothko knew his last paintings were unlike anything he had ever done. But Rothko was guarded even with Goldwater: "In his comments, fragmentary, brief, punctuated with long and heavy silences, and in his questions, freighted with suppressed intensity, meanings were never mentioned."[100] Goldwater traced Rothko's descent to the lower register of color and to the solemn darkness of the last mural cycle, concluding that "the sense of the tragic" becomes increasingly dominant. The landscape implications of the stressed horizon line, according to Goldwater, were accepted by Rothko who was aware of the suggestion of deep space. Yet, abstraction wins out in these lonely paintings that Goldwater said "reject participation and draw into themselves."

The restrained palette in most of the last paintings—variations of gray deepened with brown, black, or ocher, and befogged whites— is related to Rothko's earlier technique of oppositions, but now the effect would be heavy, portentous, airless. The symbolic use of the mirrored shapes becomes demanding. A few years earlier Sollers could write on Rothko's more lyrical, chromatic paintings:

> It is not only the systole and diastole of the eye, contraction and dilation, but a symmetrical dialogue, rising-setting, East-West North-South, a disorientation, a re-polarization, a universal re-animation . . . The color is, so to speak, the veritable master-key of analogies, a sort of larger common denominator of a materialized communal passion: Unlike geometrical forms which push things into the back ground, it contains them. Rothko's

"living" rectangles, contradicted and juxtaposed with a dominant color create a veritable "realm."

This Heideggerean apprehension of Rothko's spaces, and his dialectical method, is basic to the whole oeuvre. No matter that in the last paintings a certain blunt opacity screens the experience. It is still an attempt to make a place, a realm, a country which, as Heidegger conceived it, is the function of the visual artist who must, from the beginning, face the eternal question: what is space? Profane spaces, he wrote, are always the privation of sacred spaces going far, far back. He asks:

> Is space one of the *Urphänomenen* [the primal phenomena] the contact with which, as Goethe thought, submerges a man—once he arrives at perceiving them—in a kind of apprehension that can even reach anguish? Because behind space, so it seems, there is nothing . . . And before it, there is no possible flight to anything else.[101]

This kind of spatial metaphysical anguish was already implicit in the large canvases that Rothko liked to call "tragic" and in which he denied the traditional functions of color. (What an irony that careless commentators still talk about his work as "color fields," a vulgar simplification that robs his achievement of its depth.) In the large works, as Andrew Forge has said, Rothko discovered that "a painting sufficiently large so that when you stand close, the edges are grayed off to one's peripheral vision, takes on a kind of presence in its surface that renders internal relationships irrelevant. The moment color and scale begin a dialogue, a close viewing range is like opening a door into an internal realm."[102] This, Forge thinks, is Rothko's grandeur, his limitation, his heaven and his hell.

The interior realm was where Rothko wished to or perhaps could only live, and what he hoped to express. The "theater of the mind," as Mallarmé called it, was immensely dramatic for Rothko. His darkness at the end did allude to the light of the theater in which, when the lights are gradually dimmed, expectation mounts urgently. Such darkness for Rothko was a crucible for imagination, even as he imagined terrible things. If his last two years were hellish, his paintings reflect them faithfully. It would be futile to see them as anything other than a mournful reckoning of his life's preoccupations, birth, dissolution, and death.

AFTERWORD

"It's a funny thing to say, but I mean for me they bring news. Which is not only in painting news. Every painting brings news—it's beyond the painting, right?"[103]

De Kooning's response to Rothko's later works was to the "news" that is beyond painting. For this Rothko would have been grateful. He was of the same generation and to some degree they had all worked with a sense of mission. If one accepts Kierkegaard's view of "the religious" aspect of a man's acts, then Rothko's enterprise was religious. He had begun life as a committed man. When he preached—for that is what he did in his public statements—he preached ethics. His Existentialist leanings eventually overcame his ambitions as a public missionary, but the *sense* of mission never left him. His view is apparent in his response to George Dennison's article "The Moral Effect of the Legend of Genet" published in 1967 in the first issue of *New American Review*. Dennison, who had frequently talked with Rothko during the late forties, had written about his work, and used to see him at long intervals in later years, writes:

> Mark said "I feel it has a bearing on our cause." I thought this was astute of him, since the aesthetics really were relevant but had been put entirely in literary and psychological terms. No one but he ever noticed their relevance to abstract art. But what was much more striking was his saying "our cause" since his colleagues then were mostly rich, famous and dispersed.[104]

In Dennison's essay it is not difficult to discern the significance for Rothko. There are phrases scattered throughout that are typical of Rothko's way of thinking, such as a quote from *The Thief's Journal:*

> Was what I wrote true? False? Only this book of love will be real. What of the facts which served as its pretext? I must be their repository. It is not they which I am restoring.

"Such statements as these," Dennison wrote, "make clear to us at once that we cannot construe the 'I' of Genet as being, in the usual sense, the author himself. The books are presented overtly as a means of 'becoming'" (recalling Rothko's frequent statements that a painting lives by companionship, etc.). Dennison continues:

> This "I" then—the maker, the creator—is the common ground, or starting point, from which the voyage of dehumanization begins. To the extent that we accept or appropriate the "you" with which Genet addresses us, it is in response to this "I" alone, for all the others are transformations, chimeras of poetry and truth. Only the deeds of the maker are real, and they are not acts of crime, of loveless love and of betrayal, but acts of language, of shaping—and they are dazzling in their beauty.

It may well be that Rothko recognized himself in such lines.

A painter seeing Rothko's posthumous exhibition was reminded of something Malraux had said about paintings that are not religious, but are the opposite of profane. Was this Rothko's "cause"? And if it were, could it ever get past the modern barrier of irony? To paint the opposite of profane in an increasingly profane world was certainly a mission, and perhaps an impossible one. Yet, de Kooning perceived the message as beyond painting. Others, in experiencing awe or transport before his paintings, attributed them to a mystic. This was a repugnant idea to Rothko since it was *this* world he aspired to characterize. Yet it is obvious that he was ambivalent, and that there was another, more alluring world to which he was always drawn. He was like a mystic in that he had an overweening private hunger for illumination, for personal enlightenment, for some direct experience—or at least the quality of that experience—with the transcendent. He was a mystic in the way Nietzsche described "a mystic soul . . . almost undecided whether it should communicate or conceal itself."

I believe it was this artful indecision, this purposeful equivocation that endowed Rothko's paintings with the power to draw us back. He heightened our attention by setting up an expectancy. When he hit his stride, as he did in the mid-fifties, he could surround us with large colored surfaces that both revealed and concealed. He knew how to stage a moment of stasis full of promise. I can remember entering the Janis Gallery and stopping in the center of the room. It was much as if I had entered a remote forest on a still day with nothing stirring, and heard, or imagined I heard, a single faint rustle somewhere. In the paintings there was always some all but invisible movement that I could never quite locate but that seemed to pervade the whole. And, just as in the forest I would stand attentively, every nerve mobilized, waiting for the next sound, seeking its source, so before Rothko's paintings I would await, or summon, an indication of source. The movement Rothko created was always hovering, respiring, pulsing, but never wholly described. He teased his viewer into a state of receptivity and inquiry. Unaccustomed juxtapositions of huge areas of color (or sometimes merely tone) challenged not only the eyes of the beholder but his entire psychological and motor being. Rothko's uncanny command of these often baffling juxtapositions and subtle movements transformed his viewer into more than a collaborator. If his soul was almost undecided whether to communicate or reveal itself, its movements were nonetheless suggestive. It was this very equivocation that gave back so much that had been banished from painting—a chance for metaphor, a chance for indeterminate feeling, a chance for mystery.

There was certainly a pronounced hermetic element in Rothko's oeuvre that excited some and irritated others. All kinds of explanations were summoned. Butor even goes so far as to suggest that Rothko's development occurred during the height of the McCarthy era, and that this played a role since Rothko "produces an oasis of light that protects him and judges everything else, a light intended to benefit only individuals, in secret." This Butor regards as an escape. But he forgets that many painters have responded to their time with immense indignation which is not at all endemic to their art. Pissarro, for instance, seethed with anger about unjust social conditions, yet construed a calm, evenly lighted world of delectation. It was in his temperament, just as the longing for a universal experience of unity was in Rothko's. When Rothko insisted on the materialistic base of his art, refusing mysticism, he defended himself against the abyss, declaring again and again that these are the visible facts of his existence that includes all adjectives—from joyful to sorrowful, from earthly to transcendent. Yet the desire to be ravished, quite as the mystics desired to be ravished, was always there.

The frequent allusion to Rothko's silences in critical response bear out the approach to his work that sees it as the opposite of profane. Religiousness and reverent silence are traditionally associated. Yet, which culture can prepare the viewer for such silence? It is certainly an anomaly in our own. Rothko rightly sensed that such work would find only very sensitive viewers and limited understanding. To understand his impulse would perhaps require some other context, far removed from the modern; a leap back into a state of mind utterly diverse from the modern. If his was a hidden dream of some sacred enclosure, it is better considered in the context of some very distant dream, such as that of the 12th-century monk Theophilus who wrote to an imaginary apprentice of the spirit of wisdom, understanding, counsel, fortitude, knowledge, and godliness:

> Animated, dearest son, by these supporting virtues, you have approached the House of God with confidence, and have adorned it with so much beauty; you have embellished the ceilings or walls with varied work in different colours and have, in some measure, shown to beholders the paradise of God, glowing with various flowers, verdant with herbs and foliage, and cherishing with crowns of varying merit the souls of the saints. You have given them cause to praise the Creator in the creature and proclaim Him wonderful in His works. For the human eye is not able to consider on what work first to fix its gaze; if it beholds the ceilings they glow like brocades; if it considers the walls they are a kind of paradise; if it regards the profusion of light from the windows, it marvels at the inestimable beauty of the glass and the infinitely rich and various workmanship . . .[105]

The problem of light obsessing Rothko can be seen better in other contexts. The light he craved was a light of revelation, quite literally, and it had to be concealed. Emanation was essential. Beneath his preoccupation were endless speculations of the kind Maurice Blanchot, à propos of Nietzsche, posed when he asks: why this imperialism of light? "Light illuminates, that is to say that light hides itself, that is its sly characteristic. Light illuminates: that which is illuminated presents itself in an immediate presence which reveals itself without revealing that which manifests it. Light effaces its traces; invisible, it renders visible."[106] The paradox implicit in Rothko's best work is that he wished also to name the light itself and not only the things it illuminated. How often in his later works there are flares of burning light, sparks that glow uncannily, sparks that seem fraught with very old associations dredged up from the mythic world in which Rothko was once immersed. These are more like the sparks in the original Book of the Zohar. Matti Megged has written that the main concerns of the Zohar, or the Book of Splendour, are the mysteries of the emanation of the divine powers of the hidden God. "On the one hand, the God himself, the Ein-Sof, is so remote from human understanding that nothing can be said about him. On the other, the whole *raison d'être* of the Kabbalah, and probably of any mysticism, is the need to grasp, by thought and imagination, the living, dynamic presence of God in the world."[107] Out of this contradiction, he writes, the author of the Zohar strived to find the language appropriate to the expression of things he knew and admitted could not be talked about:

> Here lies the tremendously powerful poetic imagination of the author. He finds words, terms, metaphors whose purpose is to express the inexpressible such as: "A spark of darkness emerged-but-not-emerged from the obscurest of obscure, from the secret of Ein-Sof, not white and not dark, not red and not green, without any color at all . . ." And from this spark, within-in-within, emerged one fountain from which all nether entities got their colors.

The Book of Zohar, then, is best received as a high order of poetry, and perhaps the same must be felt about Rothko's late works. "They are not paintings," he said.

Rothko was a man of culture, in the precise sense of culture designed by the 20th century as the opposite of, or at least different from, nature. "Culture," wrote Leopold von Wiese in 1939, "is above all not 'an order of phenomena' and is not to be found in the worlds of perceptible or conceived things. It does not belong to the world of substance; it is part of the world of values, of which it is a formal category . . . Culture is no more a thing-concept than 'plus,' 'higher,' or 'better.' " Sartre, in his summing up in "The Words," thought that "Culture doesn't save anything or anyone; it doesn't justify. But it is a product of man: he projects himself into it, he recognizes himself in it; that critical mirror alone offers him his image." In the light of both these definitions of culture Rothko was a man of culture. His mission was to go beyond the world of substances to a world, literally, of values. His experience, or the experience he deemed worthy of expression, was of things and thoughts man has made. He was making a language to cover certain kinds of experiences called passions or emotions that rise up in self-reflecting modern psyches, but that have an immemorial history in art. Rothko's painting culture embraced a wide spectrum of experiences, but all reflected his respect for the expression of the longing for transcendence: from the artists of Torcello and Ravenna to Fra Angelico, to Rembrandt, to Turner—all. Knowledge was the goal, even knowledge of the seemingly unknowable. Rembrandt remained one of Rothko's beacons—Rembrandt who knew the importance of feeling rooted in the everyday world of human emotion, and yet wished to transcend it; Rembrandt who his contemporary Joachim van Sandrart said did not hesitate to oppose or contradict the rules of art, and demanded "universal harmony." There is a late portrait in the Kimbell Art Museum in Fort Worth, of a young Jewish scholar in which Rembrandt unmistakably speaks of this tension. The young man gazes from his dark atmosphere, one eye engaged with the vision of this world, the other, shadowed, rapt in another.[108]

Rothko said he looked forward to a time when an artist would be rewarded for the meaning of his life's work. The meaning of his own life's work may emerge more clearly, or at least, differently, in the future. What it meant to many of his contemporaries is expressed in a memorial statement by Andrew Carnduff Ritchie:

> One cannot escape the feeling, avoiding all the current psychoanalytic jargon in art criticism, that there was in Rothko, raised to a pitch of poetic intensity, a Zoroastrian sense of light and darkness as symbols of goodness and evil, growing out of an inheritance from a youth spent in virgin Oregon, merging with memories of his Old Testament ancestors and a deep recall of his origins in that great land of opposites, Russia.[109]

NOTES

1. "The Romantics Were Prompted," *Possibilities,* Vol. I, Winter 1947/48.
2. Conversation with the author, January 1957.
3. Author's conversation with Moise, Rothko's older brother, 1971.
4. *Contemporary Painters* (New York: Museum of Modern Art, 1948).
5. *Essays on Art,* New York: Rudge, 1916.
6. I am indebted to Professor Rudolf Arnheim for information on Cizek.
7. "Commemorative Essay," Jan. 7, 1965, in *Milton Avery* (Greenwich, Conn.: New York Graphic Society, 1969).
8. Conversation with author, Feb. 4, 1972.
9. Interview on file at Archives of American Art.
10. "On Regionalism," *American Art, 1700-1960, Sources and Documents* (Englewood Cliffs: Prentice-Hall, 1965), p. 202.
11. Interview, 1972.
12. Gerald M. Monroe, "Art Front," *Archives of American Art Journal,* Vol. 13, No. 3, 1973.
13. Statement, catalogue for group exhibition at the David Porter Gallery, Washington, D.C., 1945.
14. Conversation with the author, 1969.
15. "Ulysses, Order and Myth," *The Dial,* November 1923.
16. *An Essay on Man* (New Haven: Yale University Press, 1944).
17. Sidney Janis, *Abstract and Surrealist Art in America* (New York: Reynal and Hitchcock, 1944).
18. *The Birth of Tragedy,* translated by Walter Kaufmann (New York: Vintage, 1967).
19. *The Portable Nietzsche* (New York: Viking, 1954).
20. *The Tiger's Eye,* October 1949.
21. First Annual Membership Exhibition Catalogue, 16-29, Rockefeller Center, New York, 1937.
22. "Nature of Abstract Art," *Marxist Quarterly,* Jan.-March 1937.

23. Thomas B. Hess, *Barnett Newman* (New York, The Museum of Modern Art, 1971).
24. Holograph ms. in Keats-Shelley House, Piazza di Spagna, Rome.
25. "Pollock Symposium," *Art News,* April 1967.
26. Catalogue for group exhibition at the David Porter Gallery, Washington, D.C., 1945.
27. John Hultberg in *A Period of Exploration, San Francisco, 1945-50* by Mary Fuller McChesney (Oakland Museum, 1973).
28. Conversation with author, 1958.
29. *New Republic,* April 24, 1944.
30. Hess, *Barnett Newman.*
31. "Jan. 17, 1947," *Ararat,* 12, Fall, 1971.
32. *The Tiger's Eye,* Dec. 15, 1948.
33. *Barnett Newman* (New York: Abrams 1978).
34. *The Truants* (New York, Anchor/Doubleday, 1982).
35. Rilke, "First Duino Elegy."
36. "The Romantics Were Prompted," *Possibilities.*
37. *Notes of a Painter,* 1908.
38. "Looking at Life with the Eyes of a Child, 1955," in Jack D. Flam, *Matisse on Art* (New York: Dutton, 1978).
39. "Interview with Eugène Tardieu," *Echo de Paris,* May 13, 1895.
40. "The Romantics Were Prompted," *Possibilities,* I.
41. Translated by B. Frechtman (New York: Philosophical Library, 1947).
42. "Indirect Language and the Voices of Silence" in *Signs* (Chicago: Northwestern University Press, 1964), p. 44.
43. Bradford Cook, *Selected Prose Poems, Essays and Letters* (Baltimore: Johns Hopkins University Press, 1956).
44. "Eye and Mind," *The Primary of Perception* (Chicago: Northwestern University Press, 1964), p. 178.
45. *Interiors,* Vol. 110, May 1951, p. 104.
46. "Eye and Mind."
47. "Indirect Language and the Voices of Silence," p. 52.
48. Conversation with the author, January 1957.
49. Jean-Paul Sartre, *Situations II* (Paris: Gallimard, 1948).
50. "Indirect Language and the Voices of Silence."
51. Unpublished doctoral dissertation, Princeton University, 1951.
52. Lippold papers, Archives of American Art.
53. Susanne K. Langer, *Feeling and Form* (New York: Scribner, 1953).
54. Interview with Betty Parsons, Archives of American Art.
55. Lee Seldes, *The Legacy of Mark Rothko* (New York: Holt, Rinehart and Winston, 1978).
56. Conversation with the author.
57. Quoted in press release of Art Institute of Chicago.
58. DeKooning, "Two Americans in Action."
59. *Time,* March 3, 1961.
60. Homage prepared to be read to members of the National Institute of Arts and Letters, January 1971.

61. "A Lecture on Something," *It Is,* no. 4, Autumn, 1959.
62. Author's transcription, partially published in *The New York Times,* Oct. 31, 1958.
63. "Mark Rothko: Portrait of the Artist as an Angry Man," *Harper's,* Vol. 24, #1422, July 1970.
64. Giulio Carlo Argan, "Fra Angelico" (Skira, 1955; Cleveland: World Publishing Co.).
65. Ibid.
66. Conversation with the author, Winter 1959.
67. D. Ashton on "Philosophy of Stanley Kunitz," *The New York Times,* Dec. 10, 1959.
68. Letter to the author, March 9, 1959.
69. "A Symposium on How To Combine Architecture, Painting and Sculpture," *Interiors,* Vol. 10, May 1951.
70. "Mark Rothko," broadcast created by Dore Ashton for Canadian Broadcasting Co.
71. Tate Report, 1968-1970.
72. *Art Digest,* November 1954.
73. Catalogue foreword, Contemporary Arts Museum, Houston, Sept. 5, 1957.
74. *Chicago Art Institute Quarterly,* Nov. 15, 1954.
75. *Quadrum,* no. 10, 1961.
76. "Reflections on the Rothko Exhibition," *Arts,* no. 35; March 1961.
77. Conversation with the author.
78. Conversation with the author.
79. "The Rothko Chapel," *Art Journal,* Vol. XXX, no. 3, Spring, 1971.
80. Werner Jaeger, *Early Christianity and Greek Paideia* (Cambridge: Harvard University Press, 1965).
81. David Sutherland Wallace-Hadrill, *The Greek Patristic View of Nature* (Manchester University Press, 1968).
82. Kurt Weitzmann, *The Age of Spirituality* (The Metropolitan Museum of Art in association with Princeton University Press, 1980).
83. "Working with Rothko," *New American Review,* No. 12, 1971.
84. "Crippled Symmetry," RES 2, Harvard College, Autumn 1981.
85. Conversation with Motherwell, March 28, 1967.
86. Flam, *Matisse on Art.*
87. Catalogue for Rothko exhibition, Kunsthaus, Zürich, March 21-May 9, 1971.
88. All quotes from conversations with the author.
89. *Spectator,* March 7, 1970.
90. "The Rothko Chapel," *Art Journal,* Vol. XXX, No. 3, Spring 1971.
91. Canadian Broadcasting Company broadcast prepared by author.
92. *Modern Painting and the Northern Romantic Tradition* (New York: Harper and Row, 1975).
93. "The Rothko Chapel," *Art in America,* Vol. 61, no. 1, Jan.-Feb. 1973.
94. Press release from *Rothko Chapel* for Concert, April 9, 1972.
95. Letter to the author, April 29, 1982.

96. Conversation with the author.
97. Motherwell notes, April 21, 1969.
98. "The Mosques of New York, or The Art of Mark Rothko," *New World Writing,* no. 21, 1962.
99. *Art de France,* no. 4, 1964.
100. "Rothko's Black Paintings," *Art in America,* Vol. 59, no. 2, March-April 1971.
101. *L'Art et l'espace* (St. Gallen: Erker-Verlag, 1969).
102. Conversation with the author, August, 1982.
103. "Interview with Joseph Liss," *Art News,* January 1979.
104. Letter to the author, Sept. 27, 1978.
105. Translated by C. R. Dodwell, Theophilus, *The Various Arts* (London: Thomas Nelson and Sons Ltd., 1961).
106. L'Entretien infini (Paris: Gallimard, 1969).
107. Matti Megged, *The Darkened Light* (Tel Aviv: Sifryat Poalim, 1981).
108. This painting is surpassingly paradoxical. The worldly eye is becalmed, somewhat cool, while the visionary eye springs from its shadowy depths with intense, very palpable light. The two suspended realms—inner and outer, or spiritual and mundane—are pronouncedly different. Yet the picture as a whole can be perceived as harmonious.
109. *Salute to Mark Rothko,* Yale University Gallery, May 66-June 20, 1971.

SELECTED BIBLIOGRAPHY

I GENERAL AND BACKGROUND MATERIAL

Argan, Giulio Carlo. *Fra Angelico.* Geneva: Skira, 1955.

Ashton, Dore. *The New York School: A Cultural Reckoning.* New York: Viking, 1972.

――――. *A Reading of Modern Art.* Cleveland: Press of Case Western Reserve Univ., 1969.

――――. *The Unknown Shore: A View of Contemporary Art.* Boston: Little, Brown, 1962.

――――. *Yes, but . . . : A Critical Study of Philip Guston.* New York: Viking, 1976.

Blanchot, Maurice. *L'Entretien infini.* Paris: Gallimard, 1969.

Breeskin, Adelyn D. *Milton Avery.* Washington, D.C.: National Collection of Fine Arts, 1969.

Cage, John "A Lecture on Something." *It Is,* no. 4 (Autumn 1959): 73-78. (On Morton Feldman.)

Feldman, Morton. "After Modernism." *Art in America* 59 (November-December 1971): 68-77.

――――. "Crippled Symmetry." *RES* 2 (Autumn 1981): 91-103.

Flam, Jack D. *Matisse on Art.* New York: Dutton, 1978.

Geldzahler, Henry. *American Painting in the Twentieth Century.* New York: Metropolitan Museum of Art, 1965.

Haskell, Barbara. *Milton Avery.* New York: Whitney Museum, 1982.

Heidegger, Martin. *Die Kunst und der Raum; L'Art et l'espace.* St. Gallen: Erker-Verlag, 1969.

Hess, Thomas B. *Abstract Painting: Background and American Phase.* New York: Viking, 1951.

――――. *Barnett Newman.* New York: Museum of Modern Art, 1971.

Jaeger, Werner. *Early Christianity and Greek Paideia.* Cambridge, Mass.: Harvard Univ. Press, 1965.

Janis, Sidney. *Abstract and Surrealist Art in America*. New York: Reynal and Hitchcock, 1944.

MacNaughton, Mary Davis, and Lawrence Alloway. *Adolph Gottlieb: A Retrospective*. New York: Arts Publishers, 1981.

Megged, Matti. *The Darkened Light*. Tel Aviv: Sifryat Poalim, 1981.

Merleau-Ponty, Maurice. *The Primacy of Perception*. Evanston, Ill.: Northwestern Univ. Press, 1964.

―――. *Signs*. Evanston, Ill.: Northwestern Univ. Press, 1964.

O'Doherty, Brian. *American Masters: The Voice and the Myth*. New York: Random House, 1973.

O'Neill, John, ed. *Clyfford Still*. New York: Metropolitan Museum of Art, 1979.

Rilke, Rainer Maria. *Duino Elegies*. Translated by J. B. Leishman and Stephen Spender. New York: Norton, 1939.

Rosenberg, Harold. *Artworks and Packages*. New York: Horizon Press, 1969.

―――. *The De-Definition of Art*. New York: Horizon Press, 1972.

Rosenblum, Robert. *Modern Painting and the Northern Romantic Tradition: Friedrich to Rothko*. New York: Harper and Row, 1975.

Sandler, Irving. *The Triumph of American Painting: A History of Abstract Expressionism*. New York: Praeger, 1970.

Sartre, Jean-Paul. *Situations II*. Paris: Gallimard, 1948.

Seitz, William C. "Abstract Expressionist Painting in America: An Interpretation Based on the Work and Thought of Six Key Figures." Ph.D. dissertation, Princeton University, 1955.

Solman, Joseph. "The Easel Division of the WPA Federal Art Project." In *New Deal Art Projects: An Anthology of Memoirs*. Edited by Francis V. O'Connor. Washington, D.C.: Smithsonian Inst. Press, 1972.

Sylvester, David, ed. *Modern Art: From Fauvism to Abstract Expressionism*. New York: Franklin Watts, 1966.

Theophilus. *The Various Arts*. Translated and edited by C. R. Dodwell. London: Thomas Nelson, 1961.

Wallace-Hadrill, David Sutherland. *The Greek Patristic View of Nature*. Manchester: Manchester Univ. Press, 1968.

Weitzmann, Kurt. *The Age of Spirituality*. New York: Metropolitan Museum of Art, 1980.

II SELECTED EXHIBITION CATALOGS AND REVIEWS

Ashton, Dore. "Art." *Arts and Architecture* 75 (April 1958): 8, 29, 32.

―――. "Lettre de New York." *Cimaise*, ser. 5, no. 4 (March-April 1958): 30-31.

Bowness, Alan. "Absolutely Abstract." *Observer* (London), 15 October 1961, p. 27.

Bulletin of the Museum of Art, Portland [Oregon] 3 (November-December 1933): n. pag.

Crehan, Hubert. "Rothko's Wall of Light: A Show of His New Work at Chicago." *Arts Digest* 29 (November 1, 1954): 5, 19.

Fifteen Americans. New York: Museum of Modern Art, 1952.

Goldwater, Robert. "Reflections on the Rothko Exhibition." *Arts* 35 (March 1961): 42-45.

Hess, Thomas B. "Rothko: A Venetian Souvenir." *Art News* 69 (November 1970): 40-41, 72-74.

Hunter, Sam. "Diverse Modernism." *New York Times*, 14 March 1948, sec. 2, p. x8.

———. Text on Rothko for *Lipton, Rothko, Smith, and Tobey.* Venice: *XXIX Exposizione Biennale Internazionale d'Arte*, 1958.

J. L. "Three Moderns: Rothko, Gromaire and Solman." *Art News* 38 (January 20, 1940): 12.

Kainen, Jacob. "Our Expressionists." *Art Front* 3 (February 1937): 14-15.

Kuh, Katharine. "Mark Rothko." *Art Institute of Chicago Quarterly* 48 (November 15, 1954): 68.

McKinney, Donald. *Mark Rothko.* Zürich: Kunsthaus Zürich, 1971.

Mark Rothko Paintings. New York: Art of This Century, 1945.

The New American Painting. New York: Museum of Modern Art, 1959.

The New York School: The First Generation, Paintings of the 1940's and 1950's. Los Angeles: Los Angeles County Museum of Art, 1965.

O'Doherty, Brian. "Art: Rothko Panels Seen." *New York Times*, April 10, 1963, p. 36.

Perocco, Guido. *Mark Rothko.* Venice: Museo d'Arte Moderno Ca'Pesaro, 1970.

Pleynet, Marcelin. "Exposition Mark Rothko." *Tel Quel*, no. 12 (Winter 1963): 39-41.

Riley, Maude. "The Mythical Rothko and His Myths." *Arts Digest* 19 (January 15, 1945): 15.

Ritchie, Andrew Carnduff. *Salute to Mark Rothko.* New Haven: Yale University Art Gallery, 1971.

Rosenblum, Robert. *Mark Rothko: The Surrealist Years.* New York: Pace Gallery, 1981.

Russell, John. "A Grand Achievement of the Fifties." *Sunday Times* (London), October 15, 1961, p. 39.

Selz, Peter. *Mark Rothko.* New York: Museum of Modern Art, 1961.

Waldemar-Geroge. *The Ten.* Paris: Galerie Bonaparte, 1936.

Waldman, Diane. *Mark Rothko: A Retrospective.* New York: Guggenheim Museum/Abrams, 1978.

III   ARTICLES ON ROTHKO

Alloway, Lawrence. "Notes on Rothko." *Art International* 6 (1962): 90-94.

Ashton, Dore. "Art: Lecture by Rothko." *New York Times*, October 31, 1958, p. 26.

———. "L'Automne à New York: Letter from New York." *Cimaise*, ser. 6, no. 2 (December 1958): 37-40.

———. "Mark Rothko." *Arts and Architecture* 74 (August 1957): 8, 31.

———. "Oranges and Lemons, An Adjustment." *Arts Magazine* 51 (February 1977): 142.

———. "The Rothko Chapel in Houston." *Studio* 181 (June 1971): 272-75.

Butor, Michel. "The Mosques of New York, or The Art of Mark Rothko." In *New World Writing*, no. 21. Philadelphia: Lippincott, 1962.

De Kooning, Elaine. "Two Americans in Action: Franz Kline and Mark Rothko." *Art News Annual* 27 (1958): 86-97, 174-179.

De Menil, Dominique (Mrs. John de Menil). "Address given in the Rothko Chapel, 26 February 1971." Transcript on deposit in Rothko file, Whitney Museum, New York.

Dennison, George. "The Painting of Mark Rothko." Unpublished, n.d. On deposit in Rothko file, Museum of Modern Art, New York.

Drudi, Gabriella. "Mark Rothko." *Appia* 2 (January 1960): n. pag.

Edwards, Roy, and Ralph Pomeroy. "Working with Rothko." In *New American Review*, no. 12. New York: Simon and Schuster, 1971.

Fischer, John. "Mark Rothko: Portrait of the Artist as an Angry Man." *Harper's*, July 1970, pp. 16-23.

Goldwater, Robert. "Rothko's Black Paintings." *Art in America* 59 (March-April 1971): 58-63.

Liss, Joseph. "Portrait by Rothko." Unpublished, n.d. On deposit in Rothko file, Whitney Museum, New York.

MacAgy, Douglas. "Mark Rothko." *Magazine of Art* 42 (January 1949): 20-21.

O'Doherty, Brian. "Rothko." *Art International* 14 (October 20, 1970): 30-44.

———. "The Rothko Chapel." *Art in America* 61 (January-February 1973): 14-18, 20.

Oeri, Georgine. "Mark Rothko." *Quadrum*, no. 10 (1961): 65-74.

———. "Tobey and Rothko." *Baltimore Museum of Art News Quarterly* 23 (Winter 1960): 2-8.

Putnam, Wallace. "Mark Rothko Told Me." *Arts Magazine* 48 (April 1974): 44-45.

Sollers, Philippe. "Le mur du sens." *Art de France* 4 (1964): 239-251.

IV ARTICLES AND STATEMENTS BY ROTHKO
(ARRANGED CHRONOLOGICALLY)

[Statement.] In *The Ten: Whitney Dissenters*. New York: Mercury Galleries, 1938. (With Bernard Bradden.)

[Letter.] In Edward Alden Jewell, "The Realm of Art: A New Platform and Other Matters: 'Globalism' Pops into View." *New York Times*, June 13, 1943, p. x9. (With Adolph Gottlieb and probable collaboration of Barnett Newman.)

"The Portrait and the Modern Artist." WNYC radio broadcast "Art in New York," October 13, 1943. (With Adolph Gottlieb; excerpts from transcript in *The New York School*, Los Angeles County Museum of Art, 1965.)

"Personal Statement." In *A Painting Prophecy—1950*. Washington, D.C.: David Porter Gallery, 1945.

"Clyfford Still." In *Clyfford Still*. New York: Art of This Century, 1946.

"The Ides of Art: 'The Attitudes of 10 Artists on Their Art and Contemporaneousness." *The Tiger's Eye* (December 1947): 44.

206

"The Romantics Were Prompted." *Possibilities* 1 (Winter 1947-48): 84.
"Statement on His Attitude in Painting." *The Tiger's Eye* (October 1949): 114.
"A Symposium on How To Combine Architecture, Painting and Sculpture." *Interiors* 110 (May 1951): 104.
Unpublished letter to Lloyd Goodrich, Director, Whitney Museum of American Art, New York. December 20, 1952. On deposit in Rothko file, Whitney Museum.
Unpublished letter to Rosalind Irvine, Whitney Museum of American Art, New York, April 9, 1957. On deposit in Rothko file, Whitney Museum.
"Editor's Letters." *Art News* 56 (December 1957): 6.
[Interview with Rothko.] In Selden Rodman, *Conversations with Artists*. New York: Devin-Adair, 1957.
[Lecture.] Delivered Fall 1958, Pratt Institute, Brooklyn. Excerpts from transcript in *The New York School*, Los Angeles County Museum of Art.
Eulogy for Milton Avery. Delivered January 7, 1965, New York Society for Ethical Culture, New York. Transcript published in Breeskin, *Milton Avery*.

ILLUSTRATIONS

8. *The Omen of the Eagle*   1942
oil on canvas 25¾ x 17¾ in.
Photo Quesada/Burke, courtesy The Mark Rothko Foundation, New York

9. *Subway (Subterranean Fantasy)*   c.1938
oil on canvas 34⁵⁄₁₆ x 46½ in.
Photo Quesada/Burke, courtesy The Mark Rothko Foundation, New York

10. Untitled   c.1936
oil on canvas 32 x 42 in.
© Estate of Mark Rothko
Photo courtesy The Solomon R. Guggenheim Museum, New York

11. *Subway Scene*   1938
oil on canvas 35 x 47¼ in.
© Estate of Mark Rothko
Photo courtesy Dr. Kate Rothko Prizel

12. Untitled   c.1936
oil on canvas 16¹⁄₁₆ x 20¹⁄₁₆ in.
Courtesy The Mark Rothko Foundation, New York

13. *Antigone*   1938-41
oil on canvas 34⁵⁄₁₆ x 46¼ in.
Courtesy The Mark Rothko Foundation, New York

14. JOAN MIRO
The Hunter (Catalan Landscape)   1923-24
oil on canvas 25½ x 39½ in.
Courtesy Collection, The Museum of Modern Art, New York. Purchase

15. *Horizontal Phantom*   1943
oil on canvas 35¾ x 47¾ in.
© Estate of Mark Rothko
Photo courtesy The Pace Gallery, New York

16. *The Syrian Bull*   1943
oil on canvas 39½ x 27½ in.
Photo David Preston, courtesy Collection Annalee Newman

17. *Hierarchical Birds*   c.1944
oil on canvas 39⅝ x 31⅝ in.
Photo Quesada/Burke, courtesy The Mark Rothko Foundation, New York

18 *Poised Elements*   1944
oil on canvas 36 x 48 in.
© Estate of Mark Rothko
Photo courtesy The Pace Gallery, New York

19. *Olympian Play*   c.1944
oil on canvas 19⅝ x 27⁹⁄₁₆ in.
Courtesy The Mark Rothko Foundation, New York

20. Untitled   1945
    oil on canvas 39¼ x 27¼ in.
    © Estate of Mark Rothko
    Photo courtesy The Pace Gallery, New York

21. Untitled   1949
    watercolor, tempera on paper 39⅞ x 25⅞ in., image; 40¹⁄₁₆ x 26⅝ in., sheet
    Photo Quesada/Burke, courtesy The Mark Rothko Foundation, New York

22. Untitled   c. 1946
    oil on canvas 39⅜ x 27⁹⁄₁₆ in.
    Courtesy The Mark Rothko Foundation, New York

23. Mark Rothko at Betty Parsons Gallery, New York, 1949
    Photo courtesy Aaron Siskind

24. *Number 22*   1949
    oil on canvas 117 x 107⅛ in.
    Photo Geoffrey Clements, courtesy Collection, The Museum of Modern
    Art, New York. Gift of the artist

25. *Number 27*   1954
    oil on canvas 81 x 86⅝ in.
    Courtesy Collection Anne W. Sowell, Fort Worth

26. *Number 9*   1958
    oil on canvas 99 x 82 in.
    Courtesy Collection Mr. and Mrs. Donald Blinken

27. Untitled   c.1944-46
    watercolor, ink on paper 22⅛ x 30¹³⁄₁₆ in. image; 22¾ x 31⁵⁄₁₆ in., sheet
    Photo Quesada/Burke, courtesy The Mark Rothko Foundation, New York

28. *Triptych* from the Harvard Murals   1962
    oil on canvas left panel 104⅞ x 117 in.; central panel 104⅞ x 180½ in.;
    right panel 104⅞ x 96 in.
    Courtesy the President and Fellows of Harvard College

29. Mark Rothko on his birthday in 1960
    © Estate of Mark Rothko
    Photo Regina Bogat, courtesy The Pace Gallery, New York

30. Rothko, Carlo Battaglia, Christopher, and Mell in Rome, 1966
    Photo courtesy Carla Panicali

1. *Interior*   c.1932
   oil on masonite 23$^{15}\!/_{16}$ x 18$^{5}\!/_{16}$ in.
   Courtesy The Mark Rothko Foundation, New York

2. *Birth of Cephalopods*   1944
   oil on canvas 39$^{5}\!/_{8}$ x 53$^{11}\!/_{16}$ in.
   Courtesy The Mark Rothko Foundation, New York

3. *Slow Swirl by the Edge of the Sea*   1944
   oil on canvas 75$^{3}\!/_{8}$ x 84$^{3}\!/_{4}$ in.
   Courtesy Collection, The Museum of Modern Art, New York. Bequest of
   Mrs. Mark Rothko

4. *Number 18*   1948-49
   oil on canvas 67$^{1}\!/_{4}$ x 55$^{7}\!/_{8}$ in.
   Courtesy Vassar College Art Gallery, Poughkeepsie, New York.
   Gift of Blanchette Hooker Rockefeller '31

5. *Number 11*   1949
   oil on canvas 68$^{1}\!/_{8}$ x 43$^{5}\!/_{16}$ in.
   Courtesy The Mark Rothko Foundation, New York

6. *Number 61*   1953
   oil on canvas 116$^{1}\!/_{2}$ x 92 in.
   Courtesy Collection Panza di Biumo, Milan

7. *Number 18*   1951
   oil on canvas 81$^{3}\!/_{4}$ x 67 in.
   Courtesy Munson-Williams-Proctor Institute, Utica, New York

8. *Blue over Orange*   1956
   oil on canvas 86 x 79 in.
   Courtesy Collection Mr. and Mrs. Donald Blinken

9. The Rothko Room
   Courtesy The Tate Gallery, London

10. Untitled   1958
    oil on canvas 105 x 149 in.
    © Estate of Mark Rothko
    Photo courtesy The Pace Gallery, New York

11. *Number 117*   1961
    oil on canvas 93 x 81 in.
    Courtesy Collection Mr. and Mrs. Donald Blinken

12. Untitled   1969
   acrylic on canvas 69⅝ x 62³⁄₁₆ in.
   Courtesy The Mark Rothko Foundation, New York

13. Study for Seagram Murals   c.1958
   gouache on paper 4¼ x 16⅞ in.
   Courtesy The Mark Rothko Foundation, New York

14. *Number 17*   1947
   oil on canvas 48 x 35⅞ in.
   Courtesy The Mark Rothko Foundation, New York

15. *Greyed Olive Green, Red on Maroon*   1961
   oil on canvas 101¹³⁄₁₆ x 89⅝ in.
   Courtesy The Mark Rothko Foundation, New York

16. *Triptych*   1965-67 The Rothko Chapel, Houston, Texas
   oil on canvas left panel 96 x 180 in.; central panel 105 x 180 in.; right panel 96 x 180 in.
   Photo Hickey & Robertson, courtesy The Rothko Chapel, Houston, Texas

BRIEF CHRONOLOGY

1903 Born Marcus Rothkowitz, September 25, in Dvinsk, Russia.

1913 Settles in the United States with mother and sister in Portland, Oregon, August 17.

1921 Graduates from high school and wins scholarship to Yale University, where he remains until 1923. Moves to New York.

1924 Enrolls at the Art Students' League, January 1924. Returns to Portland where he joins a theater company.

1925 Returns to New York and enrolls in Max Weber's classes at the League; remains a member until 1929.

1928 First exhibition in a group show at the Opportunity Galleries, New York, November 15–December 12. Becomes friendly with Milton Avery.

1929 Takes part-time job teaching children at the Center Academy, Brooklyn Jewish Center; remains on faculty until 1952. Meets Adolph Gottlieb.

1932 Marries Edith Sachar.

1933 First one-man exhibition at the Museum of Art, Portland, Oregon, of his own drawings and watercolors and those of his pupils, summer. First one-man show in New York at the Contemporary Arts Gallery, November 21.

1934 Among 200 members at inauguration of Artists' Union

1935 Helps form the group "The Ten." Group's first exhibition at the Montross Gallery, New York, December 16–January 4.

1936 A founding member of American Artists' Congress. Meets Barnett Newman. Shows with "The Ten" at Galerie Bonaparte, Paris, November 10–24. Joins easel division of WPA, until 1939.

1938   Becomes United States citizen. Shows in "The Ten: Whitney Dissenters" at the Mercury Galleries, New York, November 5–26.

1940   Shows together with Marcel Gromaire and Joseph Solman at the Neumann-Willard Gallery, New York, January 8–27, and begins to sign his works Mark Rothko. Founding member of the Federation of Modern Painters and Sculptors.

1943   Third Annual Federation of Modern Painters and Sculptors Exhibition at Wildenstein Galleries, New York, June 3–26. In response to negative criticism in the *New York Times,* writes letter with Adolph Gottlieb, June 7, published June 13.

1944   Meets Mary Alice Beistle, called Mell.

1945   First one-man show at Peggy Guggenheim's Art of This Century Gallery, New York, January 9–February 4. Divorces Edith Sachar and marries Mell. Included in "A Painting Prophecy, 1950" at the David Porter Gallery, Washington, D.C., in February.

1946   Writes catalogue foreword for Clyfford Still's first one-man exhibition at the Art of This Century Gallery, New York, February 12–March 7. Exhibits watercolors at the Mortimer Brandt Gallery, New York, April 22–May 4. Spends summer in East Hampton. Shows oils and watercolors in one-man exhibition at the San Francisco Museum, August 16–September 8. Part of show travels to Santa Barbara Museum. Becomes friendly with Robert Motherwell.

1947   One-man show at the Betty Parsons Gallery, New York. Shows annually there until 1952. Visiting instructor at the California School of Fine Arts, June 23–August 1.

1948   With Clyfford Still, Robert Motherwell, William Baziotes, and David Hare, founds the school called "Subjects of the Artist."

1949   Returns to the California School of Fine Arts as guest instructor, July 5–August 12. Included in "The Intrasubjectives" at the Samuel M. Kootz Gallery, September 15–October 3.

1950   Spring trip to England, France, and Italy. Joins seventeen other painters and ten sculptors in protest letter to the Metropolitan Museum of Art (dated May 20). A photograph published in *Life* magazine the following January earns the group the title "The Irascibles." Daughter Kathy Lynn (Kate) born December 30.

1951   Appointed Assistant Professor, Department of Design, Brooklyn College. Remains until June 1954.

1952   Included in "Fifteen Americans" exhibition at the Museum of Modern Art, New York, March 25–June 11, organized by Dorothy C. Miller.

1954   One-man exhibition at the Art Institute of Chicago, October 18–December 31, later shown in part at the Rhode Island School of Design.

1955   First of two one-man shows at the Sidney Janis Gallery, New York, April 11–May 14.

1957 Visiting artist, Tulane University, New Orleans, February–March.

1958 One of four Americans represented with one-man shows at the XXIX Biennale in Venice, June 14–October 19. Commissioned by architect Philip Johnson to paint murals for Seagram Building. Travels to Italy, France, Belgium, the Netherlands, and England. Lecture at Pratt Institute, Brooklyn in October, reported by Dore Ashton in the *New York Times* and in French magazine *Cimaise*. Refuses Guggenheim award for American section in Guggenheim International Award exhibition, New York, October 22–February 23.

1960 One-man exhibition at the Phillips Collection, Washington, D.C., May 4–31.

1961 One-man exhibition at the Museum of Modern Art, New York, January 12–March 12. Exhibition travels until 1963 to London, Amsterdam, Basel, Rome, and Paris. Receives commission from the Society of Fellows, Harvard University, for murals for the Holyoke Center.

1963 Exhibition of "Five Mural Panels Executed for Harvard University by Mark Rothko" at the Solomon R. Guggenheim Museum, New York, April 9–June 2. Son Christopher Hall born August 31.

1964 Receives commission from Dominique and John de Menil for murals for Catholic Chapel in Houston, later changed to an ecumenical chapel. Spends summer in Amagansett, Long Island.

1965 Awarded the Brandeis University Creative Arts Medal, March 28.

1966 Travels with wife and two children to Italy, France, the Netherlands, Belgium, and England, spring.

1967 Teaches summer session at the University of California in Berkeley.

1968 Stricken with a serious aneurysm of the aorta. Becomes member of the National Institute of Arts and Letters, May 28. Spends summer in Provincetown, Massachusetts.

1969 Incorporates the Mark Rothko Foundation in June. Receives Doctor of Fine Arts honorary degree at Yale University, June 9. Donates nine works originally intended for the Seagram Building to the Tate Gallery, London, stipulating that the paintings always be shown in a room of their own.

1970 Commits suicide February 25. Rothko room at the Tate Gallery opens May 29.

1971 The Rothko Chapel, Houston, Texas, dedicated February 27–28.

1978 Retrospective exhibition at the Solomon R. Guggenheim Museum, New York, October 27–January 14, 1979; later at The Museum of Fine Arts, Houston, Walter Art Center, Minneapolis, and Los Angeles County Museum of Art.

217

INDEX

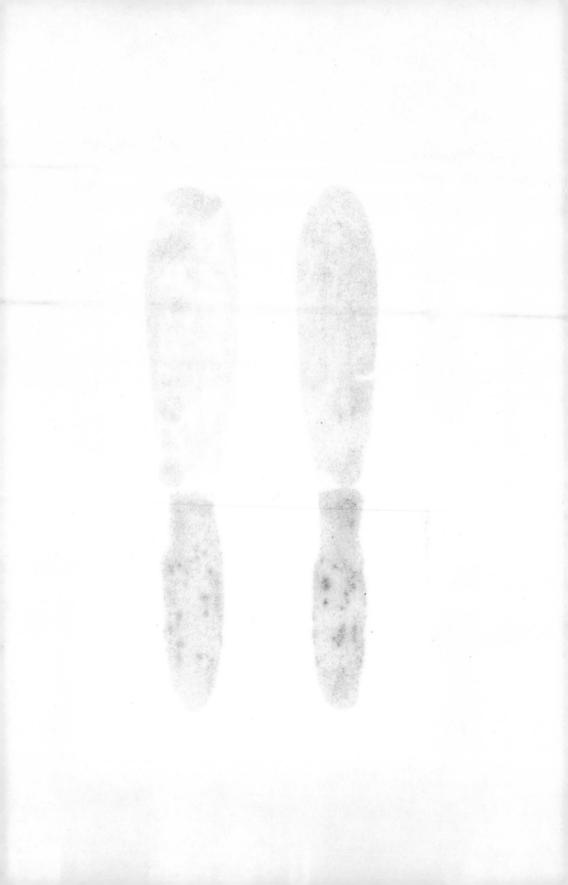

DATE DUE
